Cambridge Tracts in Mathematics
and Mathematical Physics

No. 46

LIE GROUPS

LIE GROUPS

BY

P. M. COHN

Professor of Mathematics
at Bedford College, London

CAMBRIDGE
AT THE UNIVERSITY PRESS
1968

CAMBRIDGE UNIVERSITY PRESS
Cambridge, New York, Melbourne, Madrid, Cape Town, Singapore, São Paulo, Delhi

Cambridge University Press
The Edinburgh Building, Cambridge CB2 8RU, UK

Published in the United States of America by Cambridge University Press, New York

www.cambridge.org
Information on this title: www.cambridge.org/9780521046831

First published 1957
Reprinted 1961, 1965, 1968
This digitally printed version 2008

A catalogue record for this publication is available from the British Library

ISBN 978-0-521-04683-1 hardback
ISBN 978-0-521-09298-2 paperback

PREFACE

The theory of Lie groups rests on three pillars: analysis, topology and algebra. Correspondingly it is possible to distinguish several phases, overlapping in some degree, in its development. It also allows one to regard the subject from different points of view, and it is the algebraic standpoint which has been chosen in this tract as the most suitable one for a first introduction to the subject.

The aim has been to develop the beginnings of the theory of Lie groups, especially the fundamental theorems of Lie relating the group to its infinitesimal generators (the Lie algebra); this account occupies the first five chapters. Next to Lie's theorems in importance come the basic properties of subgroups and homomorphisms, and they form the content of Chapter VI. The final chapter, on the universal covering group, could perhaps be most easily dispensed with, but, it is hoped, justifies its existence by bringing back into circulation Schreier's elegant method of constructing covering groups.

Of course whatever outlook is adopted, it is necessary to have a number of tools at one's disposal, and these have been provided in the book as far as possible. Thus before we come to Lie groups proper, the notions of analytic manifold and topological group are introduced. Lie algebras and exterior algebras are brought in later as they are needed, while theorems from analysis, such as the existence theorem for the solutions of total differential equations and the implicit function theorem, are proved in an appendix. It has been assumed that the reader has some knowledge of algebra and topology, but this need only include the elementary properties of groups and vector spaces, and the elementary notions of analytic topology.

This book owes a great deal to my colleagues in Manchester; when I gave a course on the subject in 1954, their comments showed me how much I had still to learn, and I had some opportunity of doing so in subsequent discussions with them. In particular, Dr Graham Higman and Mr G. E. H. Reuter, with their advice and comments on the earlier parts of the manuscript, saved me from a number of errors. Dr J. A. Green read the whole manuscript and made many valuable suggestions, and Dr P. J. Hilton read parts of the manuscript including the last chapter, which was much improved as a result. To all of them I should like to express my gratitude.

P.M.C.

MANCHESTER
September 1956

CONTENTS

INTRODUCTION

The theory of *continuous groups*, or as they are now called, *Lie groups*, was developed by Sophus Lie (1842–99) in connexion with the integration of systems of differential equations. These groups first arose as groups of transformations, while now they are considered just as groups in the abstract. This is similar to the situation in algebra where groups themselves appeared as permutation groups before they came to be regarded as abstract groups.

We can see how such a transformation group arises by considering the system of differential equations

$$\frac{dx_i}{dt} = u_i(x) \quad (i = 1, \ldots, n), \tag{1}$$

where x_i are the Cartesian coordinates of a point x in real Euclidean n-space. Taking $n = 3$, we can interpret the system (1) as follows: We have a fluid moving in space and the velocity of the particle of fluid at the point $x = (x_i)$ has the components $u_i(x)$. Let us consider the particle of fluid which at time $t = 0$ is at x and ask for its position at some subsequent time $t > 0$. The answer is obtained by integrating the equations (1), and is of the form

$$x_i' = f_i(x, t). \tag{2}$$

We can think of (2) as defining a transformation of the whole of space; with each point x we associate the point x' which is reached by the fluid initially at x after a time t. We express this by writing

$$x' = xS_t;$$

thus (S_t) is a family of transformations of space, and it is easy to see that these transformations form a group (if we assume that the equations (1) can be integrated for all x):

$$xS_tS_{t'} = xS_{t+t'}, \quad xS_0 = x, \quad xS_tS_{-t} = x \quad \text{(for any point } x\text{)}.$$

This group of transformations, regarded as an abstract group, is isomorphic to the additive group of real numbers.†

† Provided that the fluid never returns to its initial state (see 2.9, in particular Theorem 2.9.3).

But the way in which it arose emphasizes some special features of it:

(i) The suffix of the transformation which corresponds to the product $S_t S_{t'}$, is given by the function

$$\phi(t,t') = t + t'$$

of two variables which is continuous and even differentiable.

(ii) If we write (1) in the form

$$\delta x_i = u_i(x)\,\delta t,$$

we can consider it as an 'infinitesimal transformation'

$$x_i \to x_i + \delta x_i, \tag{3}$$

and we may think of each transformation S_t as built up by iterating the infinitesimal transformation (3). This shows that in some sense our group may be thought of as the analogue of a cyclic group, with (3) as its (infinitesimal) generator.

(iii) In order to obtain a group of transformations we had to postulate that the equations (1) have a solution for all values of x. In general, this may not be so, and we then obtain, not a group of transformations, but only a portion of a group.

This indicates the features which we shall expect of a Lie group. Thus it must be possible to introduce a coordinate system into the group, or at least part of it, such that the multiplication law of the group is expressed in terms of this coordinate system by differentiable functions. The most convenient way of taking account of the fact that this coordinate system may not be everywhere defined is to postulate that our group is a topological space in which the group operations are continuous, i.e. a *topological group*, and that a coordinate system is defined on some neighbourhood of the unit element. One can then show that although there need not be a coordinate system for the whole group, at least there is one defined on a neighbourhood of each point.

In a topological group, as in every topological space, one can distinguish between local properties, such as being locally compact or locally connected, and global properties which refer to the space as a whole. Moreover, in a Lie group one can go beyond the local to the infinitesimal. Thus one can show that a Lie group of transformations is generated by a finite number of infinitesimal transformations. In case the Lie group is given abstractly, we simply regard it as a transformation group by letting it act on itself by right translations. If the multiplication law is given by n differentiable functions $\phi_i(x, y)$, representing the coordinates of the product xy, then the general infinitesimal right translation is obtained by expanding the ϕ's in a Taylor series with respect to the y's and neglecting powers higher than the first.

To consider infinitesimal transformations instead of the finite transformations is to linearize the problem. Thus to the product of two elements of the group corresponds the sum of two infinitesimal transformations; so the latter form a linear space. Another operation is obtained by considering the commutator $x^{-1}y^{-1}xy$ of two elements of the group. This corresponds to a kind of product of infinitesimal transformations, and relative to this product the infinitesimal transformations form a non-associative linear algebra which is known as the *Lie algebra* of the group. This algebra is perhaps most familiar in the case of the group of rotations in 3-dimensional Euclidean space. It is well known that the infinitesimal rotations in space can be represented by vectors, and with this convention the usual vector product is just the multiplication for which the vectors form the Lie algebra of the rotation group.

One of the basic achievements of Lie's theory was to determine a set of conditions satisfied by the Lie algebra of a Lie group and to show that any linear algebra satisfying these conditions belongs to a Lie group.† It is this part of the theory, establishing the connexion between Lie groups and Lie algebras, together with the more fundamental properties of Lie groups, which we shall present here.

† Lie himself only considered this problem locally, i.e. he was only concerned with constructing a portion of the group near the unit element. The final step of embedding any such 'local' Lie group in a 'global' group is based on a comparatively recent theorem and belongs more to the theory of Lie algebras. For this reason a proof of this result has not been included (see Chapters VI, VII).

CHAPTER I

ANALYTIC MANIFOLDS

1.1. Charts and coordinates. One of the basic concepts to be used is that of the n-dimensional real Euclidean space. We denote this space by R^n, and indicate the coordinates of a point by attaching superscripts. Thus if $x \in R^n$, the coordinates of x are written $x^1, x^2, \ldots, x^n$, or more briefly x^i $(i = 1, \ldots, n)$. By means of the usual metric on R^n

$$d(x, y) = \{\sum_i (x^i - y^i)^2\}^{\frac{1}{2}},$$

a topology is defined on R^n, which allows us to regard it as a Hausdorff space.†

More generally, we shall consider spaces which behave locally like R^n. Thus consider a topological space T and let W be a non-empty open subspace of T which is homeomorphic to an open subspace X of R^n. If $\sigma: p \to p^\sigma$ denotes a homeomorphism of W onto X, we call σ a *chart in* T, or, more precisely, *on* W. In a given chart σ on W, each point p of W corresponds to a point $x = p^\sigma$ of R^n so that p may be described by x^i, the coordinates of x. The numbers x^i are called the *coordinates of* p (in the chart σ) and n is the *dimension* of the chart.

Now suppose that there is a homeomorphism Φ of X onto another open subspace Y of R^n and let Ψ be its inverse. If $y = x^\Phi$ is the general point of Y, with coordinates y^i $(i = 1, \ldots, n)$, then Φ and Ψ may be described by means of continuous functions ϕ^i and ψ^i:

$$\left.\begin{aligned} y^i &= \phi^i(x^1, \ldots, x^n) \\ x^i &= \psi^i(y^1, \ldots, y^n) \end{aligned}\right\} \quad (i = 1, \ldots, n).$$

Occasionally we shall use the letter x to denote the set of coordinates $(x^1, \ldots, x^n)$ as well as the point of R^n which they represent. Then the above equations of transformation between x and y may be written

$$\left.\begin{aligned} y^i &= \phi^i(x) \quad (x \in X), \\ x^i &= \psi^i(y) \quad (y \in Y). \end{aligned}\right\} \tag{1}$$

† For the topological concepts used, see, for example, Bourbaki[1].

If we combine the mappings σ and Φ from W to Y, we obtain the homeomorphism $\sigma\Phi$: $p \to (p^\sigma)^\Phi$ of W onto Y, so that $\sigma\Phi$ is again a chart on W.

Conversely, if σ and τ are any two charts on the same subspace W of T, mapping W into X and Y respectively, then $\Phi = \sigma^{-1}\tau$ is a homeomorphism of X onto Y with inverse $\Psi = \tau^{-1}\sigma$, so that the coordinates x and y of corresponding points in X and Y are related by equations of the form (1). We may regard the passage from x to y as a change of coordinates, and what has been said shows that the equations (1) (with continuous functions ϕ^i and ψ^i) are the most general equations describing a change of coordinates.†

Two charts in T whose coordinates are related by the equations (1) are said to be *analytically related* at a point p of T, if they are defined on a neighbourhood‡ of p, and if the functions ϕ^i, ψ^i occurring in (1) are analytic functions of their arguments at p^σ and p^τ respectively. Here a function $f(x)$ is said to be *analytic* at the point a of R^n if it can be expressed as a convergent power series in $x^i - a^i$ $(i = 1, \ldots, n)$ in some neighbourhood of the point a. If two charts are analytically related at every point of T at which both are defined, we say that they are *analytically related*. This is true in particular if there is no point at which both charts are defined.

1.2. Analytic structures. A topological space T is said to be *locally Euclidean* at a point p, if there exists a chart σ on a neighbourhood of p; we then say that σ is a chart *at* p. A Hausdorff space which is locally Euclidean at each point is called a *manifold*. Thus in a manifold M each point has a chart defined on some neighbourhood, a property which may be expressed by saying that the family of all charts in M *covers* M.

† It follows from the theorem on the invariance of the dimension that two charts on the same set have the same dimension (see, for example, Hurewicz, W. and Wallman, H., *Dimension Theory*, Princeton, 1941). For the particular case with which we are concerned—that of analytically related charts—this will be proved directly in 1.4.

‡ We use the term 'neighbourhood' in the sense of Bourbaki: A neighbourhood of a point p in a topological space T is a subset of T which contains p in its interior.

In this definition *all* the charts in M were admitted. We now restrict the class of charts in order to obtain a more specific structure.

DEFINITION. Let M be a Hausdorff space. Then an *analytic structure* on M is a family $\mathscr{F}$ of charts defined in M such that

M. 1. *At each point of M there is a chart which belongs to $\mathscr{F}$.*

M. 2. *Any two charts of $\mathscr{F}$ are analytically related.*

M. 3. *Any chart in M which is analytically related to every chart of $\mathscr{F}$ itself belongs to $\mathscr{F}$.*

We shall express M. 2 and M. 3 by saying that $\mathscr{F}$ is *analytic* and *maximal*, respectively. Thus an analytic structure on M is a maximal analytic family of charts covering M. It is clear that a Hausdorff space with an analytic structure is necessarily a manifold, and the space, together with this structure, is called an *analytic manifold*. By a chart in an analytic manifold we always understand a chart belonging to the analytic structure. When we wish to stress this fact we refer to the members of the structure as *admissible* charts.

In practice it is usually impossible to obtain a maximal analytic family covering a space by an explicit construction. This difficulty is overcome by the following theorem which shows that it is sufficient to construct any analytic family covering the space.

THEOREM 1.2.1. *Let M be a Hausdorff space and $\mathscr{C}$ an analytic family of charts which covers M. Then there is a uniquely determined maximal analytic family of charts which contains $\mathscr{C}$.*

Proof. Let $\mathscr{F}$ be the set of all charts in M which are analytically related to each member of $\mathscr{C}$; we shall show that $\mathscr{F}$ has the required properties. Let us express the fact that two charts σ and τ are analytically related at p by writing $\sigma \underset{p}{\sim} \tau$. If ρ, σ and τ are any charts at p, then it is easily verified that $\rho \underset{p}{\sim} \rho$, that $\rho \underset{p}{\sim} \sigma$ implies $\sigma \underset{p}{\sim} \rho$, and that $\rho \underset{p}{\sim} \sigma$, $\sigma \underset{p}{\sim} \tau$ imply $\rho \underset{p}{\sim} \tau$. Thus '$\underset{p}{\sim}$' is an equivalence relation on the set of charts at p. Now let ρ_1 and ρ_2 be any members of $\mathscr{F}$ and let W be the intersection of the sets on which ρ_1 and ρ_2 are defined. As an intersection of open sets W is again open. If W is empty, then ρ_1 and ρ_2 are analytically related by definition. Otherwise let p be any point of W; since W

is open it is a neighbourhood of p, and since $\mathscr{C}$ covers M, there is a chart σ at p which belongs to $\mathscr{C}$. By definition of $\mathscr{F}$, $\rho_1 \underset{p}{\sim} \sigma$ and $\rho_2 \underset{p}{\sim} \sigma$, whence $\rho_1 \underset{p}{\sim} \rho_2$. Thus ρ_1 and ρ_2 are analytically related at each point p of W, and hence they are analytically related. This proves that $\mathscr{F}$ is analytic. Clearly $\mathscr{F} \supseteq \mathscr{C}$, and if τ is analytically related to each member of $\mathscr{F}$ then it is analytically related to each member of $\mathscr{C}$ and hence belongs to $\mathscr{F}$. Thus $\mathscr{F}$ is a maximal analytic family containing $\mathscr{C}$. If $\mathscr{F}_1$ is another maximal analytic family containing $\mathscr{C}$, then each member of $\mathscr{F}_1$ is analytically related to each member of $\mathscr{C}$ and therefore belongs to $\mathscr{F}$. Hence $\mathscr{F}_1 \subseteq \mathscr{F}$, and similarly $\mathscr{F} \subseteq \mathscr{F}_1$, which proves that $\mathscr{F}_1 = \mathscr{F}$. Thus $\mathscr{F}$ is unique and the proof is complete.

The family $\mathscr{F}$ in Theorem 1.2.1 covers M (since $\mathscr{C}$ does) and therefore defines an analytic structure on M. So in order to define an analytic structure on a space M it is enough to specify an analytic family of charts which covers M. Of course there may be different analytic families covering M which define the same analytic structure. The necessary and sufficient condition for this to be the case is given by the

Corollary. *Let $\mathscr{C}_1$ and $\mathscr{C}_2$ be two analytic families of charts covering a space M. Then there is a maximal analytic family containing $\mathscr{C}_1$ and $\mathscr{C}_2$ if and only if for each point p of M there is a chart of $\mathscr{C}_1$ which is analytically related at p to a chart of $\mathscr{C}_2$.*

The condition is clearly necessary. Conversely, if it is satisfied, then by the argument used to prove Theorem 1.2.1, every chart of $\mathscr{C}_1$ is analytically related to every chart of $\mathscr{C}_2$ and hence the family $\mathscr{C}$ of all charts belonging to $\mathscr{C}_1$ or $\mathscr{C}_2$ is analytic. If $\mathscr{F}$ is the maximal analytic family containing $\mathscr{C}$, then $\mathscr{F} \supseteq \mathscr{C}_1$ and $\mathscr{F} \supseteq \mathscr{C}_2$; this proves the corollary.

As an example let us consider the surface of a unit sphere S in three dimensions. At any point p on S take a great circle C through p and take a system of latitude and longitude with C as the equator and p as defining the 'Greenwich meridian'. If the poles of S with respect to C ('north and south poles') are joined by a line l not passing through p (the 'date-line'), then the complement of l is a neighbourhood of p on which latitude and longitude define a chart. If the same construction is carried out for another point q of S, then the two charts are analytically related. This is easily verified by choosing the centre of S to be at the origin (0, 0, 0), the point p at (1, 0, 0) and by observing that the Cartesian coordinates of the general point of S in terms of latitude θ and longitude ϕ at p are $(\cos\theta\cos\phi, \cos\theta\sin\phi, \sin\theta)$. By equating these expressions to the corresponding expressions in terms of the chart at q we obtain analytic relations which can be solved for either set. Thus we have an analytic family of charts covering S, and by Theorem 1.2.1 this defines an analytic structure on S.

Another analytic family covering S may be obtained as follows: We take $p \in S$ to be the north pole and consider the stereographic projection from the south pole on the plane through the equator. This is a homeomorphism of the punctured sphere (namely, the sphere with the south pole removed) and the Euclidean plane, and hence defines a chart at p. In order to obtain the coordinates of the general point q in this chart, we map it into q_0, the point in which the straight line from the south pole $(0, 0, -1)$ to q cuts the (x, y)-plane. If the Cartesian coordinates are (x, y, z), its coordinates in the chart are $\left(\frac{x}{1+z}, \frac{y}{1+z}\right)$, the plane coordinates of q_0. It is again not hard to verify that if such a chart is constructed at each point of S, then these charts form a second analytic family. Moreover, the charts in these two families are analytically related. Hence, by the corollary to Theorem 1.2.1, these two families are contained in the same maximal analytic family and therefore define the same analytic structure.

We note the following examples of analytic manifolds:

1. The space R^n. The Cartesian coordinates in R^n serve as a chart at every point. We shall denote the analytic manifold so defined by $\mathfrak{R}^n$; for $\mathfrak{R}^1$ we shall also write $\mathfrak{R}$.

2. The torus T^n. This is the subspace of the n-dimensional complex Euclidean space described by

$$z_\nu = \exp 2\pi i\theta_\nu \quad (0 \leqslant \theta_\nu < 1). \tag{2}$$

Topologically the torus is the Cartesian product of n circles. In particular, for $n=1$ we obtain a circle, and for $n=2$ the familiar anchor ring. The formula (2) can be used to define an analytic structure on T^n, and the analytic manifold so obtained is denoted by $\mathfrak{T}^n$. We also write $\mathfrak{T}$ instead of $\mathfrak{T}^1$ and sometimes refer to $\mathfrak{T}$ as the (analytic manifold of) real numbers mod 1.

3. The set $GL(n, R)$ of all automorphisms of a vector space V of dimension n over R, that is, the general linear group. In terms of a given basis of V the automorphisms may be expressed as non-singular $n \times n$ matrices with coefficients in R, and the n^2 coefficients serve as a chart at each point of $GL(n, R)$.

4. A single point, or, more generally, any discrete space,† may be regarded as a 'zero-dimensional' analytic manifold.

Ex. Show that there are distinct analytic structures on R which induce the same topology (consider the coordinate transformation $y = x^3$).

1.3. Real functions on a manifold. Let M be a manifold and f a real-valued function defined on a part (possibly the whole) of M. We shall express this by saying that f is defined *in* M. If σ is a chart on some subset W of M on which f is defined, then we can express f as a function of n real variables by writing

$$\bar{f}(x) = f(p), \tag{3}$$

where $x = p^\sigma$. If τ is another chart on W which is related to σ by (1), then we can express f similarly in terms of τ: $f(p) = \bar{\bar{f}}(y)$ $(y = p^\tau)$. It is clear that $\bar{f}$ and $\bar{\bar{f}}$ are related by the equations

$$\left.\begin{aligned} \bar{f}(x) &= \bar{\bar{f}}(\phi(x)), \\ \bar{\bar{f}}(y) &= \bar{f}(\psi(y)), \end{aligned}\right\} \tag{4}$$

which hold identically in x and y.

In the foregoing discussion, where only two charts occurred, it was more convenient to denote the coordinates of the general point p in these charts by x and y respectively instead of p^σ and p^τ. We shall adopt this practice generally and even refer to a given chart by naming the coordinate functions which it defines, rather than by its proper name. For distinction we enclose the symbol for the coordinates in brackets; thus in future we shall usually speak of charts (x), (y), ... and not σ, τ, ...

† A topological space is said to be *discrete*, if any subset consisting of a single point is open.

A real-valued function f defined in M is a mapping of one topological space into another, and so we know what it means for f to be continuous. From the definitions in 1.1 and 1.2 we see that f is continuous at a point p if and only if its expression in terms of a chart at p is a continuous function of its n arguments at p. If $\mathfrak{M}$ is an analytic manifold, we shall say that a real-valued function f in $\mathfrak{M}$ is *analytic* at a point p if it is defined on some neighbourhood of p and its expression in terms of an admissible chart σ at p is an analytic function of its arguments at p^σ. It is easily verified that this definition does not depend on the choice of the chart σ. A function which is analytic at every point at which it is defined is called *analytic*. We shall denote by $\mathscr{A}_p$ the set of analytic functions which are defined at p, and by $\mathscr{A}$ the set of all analytic functions in $\mathfrak{M}$.

Ex. 1. An analytic function of an analytic function is analytic.

Ex. 2. Each coordinate of an admissible chart is analytic.

1.4. Tangent vectors. Let $\mathfrak{M}$ be an analytic manifold, p a point of $\mathfrak{M}$ and (x) a chart at p (understood to be admissible). Suppose that we are given a certain direction at p; this may be specified by laying a smooth curve through p in the given direction, and describing the curve by a parameter t:

$$x^i = x^i(t) \equiv x_0^i + \lambda^i t + O(t^2) \tag{5}$$

for small t. Here x_0^i are the coordinates of p and the λ^i are constants which do not all vanish, provided that t is suitably chosen. By differentiating (5) we obtain

$$\left[\frac{dx^i}{dt}\right]_{t=0} = \lambda^i,$$

and these numbers λ^i define the given direction completely.† For example, given an analytic function f defined at p: $f(x)$, its derivative in the direction given by (5) is

$$\left[\frac{d}{dt} f(x(t))\right]_{t=0} = \lambda^i \left[\frac{\partial f}{\partial x^i}\right]_p. \tag{6}$$

† However, the direction depends only on the ratios of the λ's and not on the λ's themselves.

On the right the summation convention has been used, which consists in summing over the appropriate range—usually from 1 to n—with respect to any suffix (in this case i) which occurs twice. We shall use this convention throughout the text, except when otherwise stated. The subscript p in (6) indicates that the function in square brackets is to be evaluated at the point p.

The formula (6) suggests considering $L=\lambda^i.\partial/\partial x^i$ (evaluated at p) as an operator on $\mathscr{A}_p$ with real values. Accordingly we define: An operator of the form $L=\lambda^i.\partial/\partial x^i$, where the λ^i are any real constants, will be called a *tangent vector* at p. The definition involves a particular chart and we therefore give an alternative characterization of tangent vectors in

THEOREM 1.4.1. *A mapping L of $\mathscr{A}_p$ into R is a tangent vector if and only if it is linear over R:*

$$L(\alpha f+\beta g)=\alpha.Lf+\beta.Lg \quad (f,g\in\mathscr{A}_p;\ \alpha,\beta\in R), \tag{7}$$

and satisfies

$$L(fg)=Lf.g(p)+f(p).Lg \quad (f,g\in\mathscr{A}_p). \tag{8}$$

For clearly every tangent vector satisfies (7) and (8); equation (8) is just the product rule for differentiation. Now let L be a mapping of $\mathscr{A}_p$ into R which satisfies (7) and (8). Then for any constant function c in $\mathscr{A}_p$ we have

$$Lc=c.L1=c(L1.1+1.L1)=2Lc,$$

whence $Lc=0$. Now let $f\in\mathscr{A}_p$; near p we may express f as

$$f(x)=f(x_0)+c_i(x^i-x_0^i)+(x^i-x_0^i)\,(x^j-x_0^j)\,g_{ij}(x),$$

where $g_{ij}\in\mathscr{A}_p$, x_0^i are the coordinates of p and $c_i=[\partial f/\partial x^i]_p$. Applying L and using (7), we obtain

$$Lf=Lf(x_0)+c_i.L(x^i-x_0^i)+L((x^i-x_0^i)\,(x^j-x_0^j)\,g_{ij}(x)).$$

The first term vanishes because $f(x_0)$ is constant. For the last sum on the right we have, by (8),

$$\begin{aligned}L((x^i-x_0^i)\,(x^j-x_0^j)\,g_{ij}(x))&\\ =L(x^i-x_0^i)\,[(x^j-x_0^j)\,g_{ij}]_p&+L(x^j-x_0^j)\,[(x^i-x_0^i)\,g_{ij}]_p\\ &+L(g_{ij})\,[(x^i-x_0^i)\,(x^j-x_0^j)]_p=0.\end{aligned}$$

Hence

$$Lf = Lx^i . c_i = Lx^i \left[\frac{\partial f}{\partial x^i}\right]_p, \quad \text{i.e.} \quad L = \lambda^i \frac{\partial}{\partial x^i}, \quad \text{where} \quad \lambda^i = Lx^i.$$

Thus L is a tangent vector, as was to be proved.

We have proved incidentally that a tangent vector L may, in any chart (x), be expressed by the formula

$$L = Lx^i \frac{\partial}{\partial x^i}. \tag{9}$$

If L_1 and L_2 are tangent vectors at p, and $\alpha_1, \alpha_2 \in R$, then $\alpha_1 L_1 + \alpha_2 L_2$, defined by $(\alpha_1 L_1 + \alpha_2 L_2) f = \alpha_1 L_1 f + \alpha_2 L_2 f$, is again a tangent vector at p; therefore the tangent vectors at p form a vector space over R. We shall denote this space by $\mathfrak{L}_p$.

Theorem 1.4.2. *If (x) is an admissible chart at p, then the tangent vectors $\partial/\partial x^i$ form a basis of the space $\mathfrak{L}_p$.*

Proof. It is clear that the operator $\partial/\partial x^i : f \to [\partial f/\partial x^i]_p$ is in $\mathfrak{L}_p$, and equation (9) shows that the $\partial/\partial x^i$ span $\mathfrak{L}_p$. To prove their independence, suppose that there is a linear relation between them, say

$$L \equiv \lambda^i \frac{\partial}{\partial x^i} = 0. \tag{10}$$

Since the jth coordinate x^j is in $\mathscr{A}_p$, we may apply L to it:

$$0 = Lx^j = \lambda^i \frac{\partial x^j}{\partial x^i} = \lambda^j.$$

This shows that all the coefficients λ^j in (10) must be zero and the theorem follows.

Suppose now that (x) and (y) are two charts at p, of dimensions m and n respectively. By Theorem 1.4.2, each of the sets $\partial/\partial x^i \, (i = 1, \ldots, m)$, $\partial/\partial y^j \, (j = 1, \ldots, n)$ is a basis of $\mathfrak{L}_p$, and hence $m = n$. This proves

Theorem 1.4.3. *All admissible charts at a given point of an analytic manifold have the same dimension.*

We may therefore define the *dimension* of an analytic manifold at a point p as the dimension of any chart at p. If a manifold has the same dimension n at all its points, it is said to be of dimension n. It is usual to require that a manifold shall have the same dimension at all its points; although we do not make this assumption explicitly, it is in fact true in all the cases which we consider.

As another application of (9) we derive the formula for a change of coordinates in $\mathfrak{L}_p$: If (y) is a second chart at p, then by (9),

$$\frac{\partial}{\partial y^k} = \left[\frac{\partial x^i}{\partial y^k}\right]_p \frac{\partial}{\partial x^i}.$$

Hence
$$\frac{\partial}{\partial y^k} = \alpha^i_k \frac{\partial}{\partial x^i}, \tag{11}$$

where $\alpha^i_k = [\partial x^i/\partial y^k]_p$ is the Jacobian matrix of the equations of transformation. Thus in the space $\mathfrak{L}_p$ of tangent vectors we can describe all coordinate changes by *linear* transformations.

Together with $\mathfrak{L}_p$ we wish to consider another vector space associated with the point p, namely, the dual space $\mathfrak{L}^*_p$ of $\mathfrak{L}_p$. For the sake of clarity we shall, in the next section, briefly review the properties of the dual space which we require.

1.5. The dual vector space. Let V be a vector space over R. *A linear form* on V is a mapping ξ of V into R: $v \to \langle v, \xi\rangle$, such that
$$\langle \alpha u + \beta v, \xi\rangle = \alpha\langle u, \xi\rangle + \beta\langle v, \xi\rangle \quad (u, v \in V;\ \alpha, \beta \in R).$$

The set of all linear forms on V will be denoted by V^*. We can define addition and multiplication by scalars in V^* as follows:

$$\langle u, \alpha\xi + \beta\eta\rangle = \alpha\langle u, \xi\rangle + \beta\langle u, \eta\rangle \quad (u \in V;\ \xi, \eta \in V^*;\ \alpha, \beta \in R).$$

It is easily seen that with these definitions V^* is a vector space over R. It is called the *dual space* of V, and $\langle u, \xi\rangle$ is called the *inner product* of u and ξ.

If $v_1, \ldots, v_n$ is a basis of V, then the general element v of V has the form
$$v = \lambda^i v_i, \tag{12}$$

and for any suffix j the mapping $v \to \lambda^j$ is a linear form on V, which thus defines an element ξ^j of V^*. With this notation the expression (12) for the general element of V becomes

$$v = \langle v, \xi^i\rangle v_i. \tag{13}$$

We shall prove that the ξ^i $(i = 1, \ldots, n)$ form a basis of V^*. For this purpose we note first that

$$\langle v_i, \xi^j\rangle = \delta^j_i, \tag{14}$$

where δ_i^j, the Kronecker delta, is defined to be 1 if $i=j$ and 0 otherwise. Equation (14) is simply a consequence of the definition of ξ^j. Next we show that

Any elements ξ^j of V^* which satisfy (14) *for some basis v_i of V, form a basis of V^*.*

For let $\eta \in V^*$; since the v_i span V, the element η is uniquely determined by its values on the v_i. In particular, the elements ξ^i are uniquely determined by (14); further, given any numbers $k_i \in R$, the element $k_j \xi^j$ has the same values as η on the v_i, if and only if $\langle v_i, \eta \rangle = \langle v_i, k_j \xi^j \rangle = k_j \langle v_i, \xi^j \rangle = k_i$. Thus η can be expressed uniquely as a linear combination of the ξ^i: $\eta = \langle v_i, \eta \rangle \xi^i$. We sum up these facts in

THEOREM 1.5.1. *If V is a vector space of dimension n over R, with the basis $v_1, \ldots, v_n$, and V^* is the dual space, then the dimension of V^* is also n, and V^* has a basis ξ^j satisfying*

$$\langle v_i, \xi^j \rangle = \delta_i^j. \tag{14}$$

This basis is uniquely determined by the equations (14) *and the general element of V can be expressed as*

$$v = \langle v, \xi^i \rangle v_i. \tag{13}$$

The basis ξ^i of V^* determined in this way is called the basis *dual* to the basis v_i of V or, when the v_i are understood, simply the *dual basis.*

If U and V are vector spaces over R with duals U^* and V^* respectively, and Φ is any linear mapping of U into V, then we can define a linear mapping Φ^* of V^* into U^* by the rule

$$\langle u, \Phi^* \eta \rangle = \langle u\Phi, \eta \rangle \quad (u \in U, \eta \in V^*).$$

The mapping Φ^* is called the *transpose* of Φ.

Ex. In terms of bases $u_1, \ldots, u_m$; $v_1, \ldots, v_n$ of U and V, any linear mapping Φ of U into V can be represented by an $m \times n$ matrix (a_{ij}): $u_i \Phi = a_{ij} v_j$. Show that the matrix of the mapping Φ^* of V^* into U^*, referred to the dual bases, is the transpose of (a_{ij}).

1.6. Differentials. Let $\mathfrak{M}$ again be an analytic manifold. We wish to determine $\mathfrak{L}_p^*$, the dual of the space $\mathfrak{L}_p$ of tangent vectors at p. For this purpose we note that each function $f \in \mathscr{A}_p$

defines a linear form on $\mathfrak{L}_p$ by the rule: $L \to Lf$. This mapping of $\mathfrak{L}_p$ into R is denoted by df; thus df is the element of $\mathfrak{L}_p^*$ defined by

$$\langle L, df \rangle = Lf.$$

In particular, if (x) is any chart at p, then the elements dx^i form a basis of $\mathfrak{L}_p^*$. For they belong to $\mathfrak{L}_p^*$ and they satisfy

$$\left\langle \frac{\partial}{\partial x^j}, dx^i \right\rangle = \left[\frac{\partial x^i}{\partial x^j}\right]_p = \delta^i_j.$$

Therefore, by Theorem 1.5.1, they form the basis of $\mathfrak{L}_p^*$ dual to the basis $\partial/\partial x^i$. Thus every element of $\mathfrak{L}_p^*$ has the form $\mu_i \, dx^i$, where $\mu_i \in R$; for example, if $f \in \mathscr{A}_p$, then $df = [\partial f/\partial x^i]_p \, dx^i$. The elements of $\mathfrak{L}_p^*$ are called *differentials* at p.

Let $z^1, \ldots, z^m$ be any elements of $\mathscr{A}_p$, and consider a function w defined by $w = \phi(z^1, \ldots, z^m)$. Then $w \in \mathscr{A}_p$, provided that ϕ is analytic for the values which the arguments $z^\mu \, (\mu = 1, \ldots, m)$ assume at p. Let us calculate dw: If (x) is a chart at p, then

$$dw = \left[\frac{\partial \phi}{\partial x^i}\right]_p dx^i = \left[\frac{\partial \phi}{\partial z^\mu}\right]_p \left[\frac{\partial z^\mu}{\partial x^i}\right]_p dx^i$$
$$= \left[\frac{\partial \phi}{\partial z^\mu}\right]_p dz^\mu.$$

Hence
$$dw = \left[\frac{\partial \phi}{\partial z^\mu}\right]_p dz^\mu. \tag{15}$$

THEOREM 1.6.1. *Let $z^1, \ldots, z^m \in \mathscr{A}_p$. Then each function in $\mathscr{A}_p$ can be expressed as an analytic function of $z^1, \ldots, z^m$ if and only if $dz^1, \ldots, dz^m$ span $\mathfrak{L}_p^*$.*

Proof. Let (x) be a chart at p and suppose that each function in $\mathscr{A}_p$ can be expressed as an analytic function of the z's. Then, in particular,
$$x^i = \psi^i(z^1, \ldots, z^m),$$
and hence, by (15)
$$dx^i = \left[\frac{\partial \psi^i}{\partial z^\mu}\right]_p dz^\mu;$$
the elements dx^i, and therefore also the elements dz^μ, span $\mathfrak{L}_p^*$. Conversely, suppose that the dz^μ span $\mathfrak{L}_p^*$; by differentiating the equations
$$z^\mu = \phi^\mu(x^1, \ldots, x^n) \tag{16}$$
(which express the fact that $z^\mu \in \mathscr{A}_p$) we obtain
$$dz^\mu = \left[\frac{\partial \phi^\mu}{\partial x^i}\right]_p dx^i,$$

and since the dz^μ span $\mathfrak{L}_p^*$, the matrix $[\partial\phi^\mu/\partial x^i]_p$ must be of rank n. Hence $n \leqslant m$ and, by the implicit function theorem (Theorem A3 of the Appendix) the equations (16) can be solved for $x^1, \ldots, x^n$ in terms of n of the z^μ:

$$x^i = \psi^i(z^{\mu_1}, \ldots, z^{\mu_n}), \tag{17}$$

where the ψ's are analytic for the values which the z^μ assume at p. Since each function of $\mathscr{A}_p$ can be expressed as an analytic function of the x^i, the theorem follows.

If in this theorem we take $m = n$, then $dz^1, \ldots, dz^n$ span $\mathfrak{L}_p^*$ if and only if they form a basis, and (16) and (17) show that the hypothesis on the z's may be expressed by saying that $z^1, \ldots, z^n$ define a chart at p. This proves

THEOREM 1.6.2. *The functions $z^1, \ldots, z^n$ in $\mathscr{A}_p$ define an admissible chart at p if and only if the differentials $dz^1, \ldots, dz^n$ form a basis of $\mathfrak{L}_p^*$.*

Ex. 1. Show that if $z^1, \ldots, z^m \in \mathscr{A}_p$ are such that there is no functional relation between them, then $dz^1, \ldots, dz^m$ are linearly independent, but that the converse is not always true.

Ex. 2. Deduce (9), in the form $L = \langle L, dx^i \rangle \, \partial/\partial x^i$, by applying Theorem 1.5.1.

1.7. Infinitesimal transformations and differential forms. In 1.4 we saw that a tangent vector at a point p of an analytic manifold $\mathfrak{M}$ is an expression of the form $\lambda^i . \partial/\partial x^i$. Here λ^i is to be thought of as specifying a direction at p, namely, the direction of the 'infinitesimal vector' with the components $\lambda^i dt$. On the other hand, a differential at p has the form $\mu_i \, dx^i$. We think of this as the mapping of $\mathfrak{L}_p$ into R defined by

$$\lambda^i dt \to \lambda^i \mu_i.$$

We now wish to consider collections of tangent vectors or of differentials at different points of $\mathfrak{M}$.

DEFINITION. An *infinitesimal transformation* X on $\mathfrak{M}$ is a collection of tangent vectors X_p, one at each point p of $\mathfrak{M}$. A *differential form* ω on $\mathfrak{M}$ is a collection of differentials ω_p, one at each point p of $\mathfrak{M}$.

An example of a differential form is the collection of differentials df, where f is an analytic function defined on the whole of $\mathfrak{M}$. This form will also be denoted by df; whether df means the differential form or its value at a particular point is usually clear from the context.

Let X be an infinitesimal transformation and let $f \in \mathscr{A}$. At each point p at which f is defined, X associates a real number with f, namely, $X_p f$. As p varies we thus obtain a real function Xf whose domain is the domain of f. We shall say that X is *analytic* if, for any $f \in \mathscr{A}$, the function Xf again belongs to $\mathscr{A}$; that is, X is analytic if and only if it maps $\mathscr{A}$ into itself. The set of all analytic infinitesimal transformations will be denoted by $\mathfrak{L}$.

THEOREM 1.7.1. *The set $\mathfrak{L}$ consists of precisely those mappings X of $\mathscr{A}$ into itself which have the following properties:*

(i) *Given any $p \in \mathfrak{M}$, the function Xf is defined at p for all $f \in \mathscr{A}_p$.*

(ii) $X(\alpha f + \beta g) = \alpha Xf + \beta Xg \qquad (f, g \in \mathscr{A};\ \alpha, \beta \in R).$

(iii) $X(fg) = Xf.g + f.Xg \qquad (f, g \in \mathscr{A}).$

Proof. If $X \in \mathfrak{L}$, then X maps $\mathscr{A}$ into itself and (i) holds by definition, while (ii) and (iii) follow from Theorem 1.4.1. Conversely, if a mapping X of $\mathscr{A}$ into itself satisfies (i)–(iii), then again by Theorem 1.4.1, the mapping $f \to [Xf]_p$, where $f \in \mathscr{A}_p$, is a tangent vector, defined for all points p of $\mathfrak{M}$. Therefore X is an infinitesimal transformation, which is analytic because it maps $\mathscr{A}$ into itself. This completes the proof.

Now let an infinitesimal transformation X and a differential form ω be given. At any point p of $\mathfrak{M}$ they define a tangent vector X_p and a differential ω_p respectively, and we can form their inner product $\langle X_p, \omega_p \rangle$, which is a real number. As p varies, we obtain a real function, defined on the whole of $\mathfrak{M}$, which we denote by $\langle X, \omega \rangle$. Its value at p will be denoted by $\langle X, \omega \rangle_p$, so that $\langle X, \omega \rangle$ may be defined by the equation

$$\langle X, \omega \rangle_p = \langle X_p, \omega_p \rangle. \tag{18}$$

If (x) is a chart in $\mathfrak{M}$, we shall sometimes use the expressions $\langle \partial/\partial x^i, \omega \rangle$ and $\langle X, dx^i \rangle$; they may be defined as in (18) and represent real functions defined on the region covered by the chart (x).

Let us now choose a chart (x) in $\mathfrak{M}$; an infinitesimal transformation X may be expressed in terms of this chart as

$$X = \xi^i(x) \frac{\partial}{\partial x^i}, \tag{19}$$

where the $\xi^i(x)$ are real functions defined by $\xi^i(x) = Xx^i$. The expression $\xi^i . \partial/\partial x^i$ is sometimes called the *symbol* for X in the given chart. Similarly a differential form ω may be expressed as

$$\omega = w_i(x)\, dx^i, \tag{20}$$

where $w_i = \langle \partial/\partial x^i, \omega \rangle$. Using the inner product, we may express (19) and (20) as follows:

$$X = \langle X, dx^i \rangle \frac{\partial}{\partial x^i}, \tag{21}$$

$$\omega = \left\langle \frac{\partial}{\partial x^i}, \omega \right\rangle dx^i. \tag{22}$$

Equation (19) suggests that we define the addition of infinitesimal transformations X and Y, and multiplication by scalars, by the rule

$$(\alpha X + \beta Y)_p = \alpha X_p + \beta Y_p \quad (\alpha, \beta \in R).$$

If X and Y are analytic, then so is $\alpha X + \beta Y$, and in this way we may regard $\mathfrak{L}$ as a vector space (in general of infinite dimension) over R.

From (21) it follows without difficulty that an infinitesimal transformation is analytic if and only if for each chart (x) in $\mathfrak{M}$, X may be written in the form $X = \xi^i . \partial/\partial x^i$, where $\xi^i \in \mathscr{A}$. In analogy with this we define a differential form ω on $\mathfrak{M}$ to be *analytic*, if for each chart (x) in $\mathfrak{M}$, ω can be expressed as

$$\omega = w_i\, dx^i$$

with coefficients $w_i \in \mathscr{A}$. As is easily seen, it is enough to require this condition to hold for some family of (admissible) charts which covers $\mathfrak{M}$.

The set of all analytic differential forms on $\mathfrak{M}$ is denoted by $\mathfrak{L}^*$. This set may be defined as a vector space over R (using the definition suggested by (20)) just in the way $\mathfrak{L}$ was defined. Of course $\mathfrak{L}^*$ will not be the dual of $\mathfrak{L}$, since its elements are not linear functions on $\mathfrak{L}$; the value $\langle X, \omega \rangle$ (for $X \in \mathfrak{L}$, $\omega \in \mathfrak{L}^*$) lies in $\mathscr{A}$, not in R.

To illustrate the notion of an infinitesimal transformation we take the Euclidean plane R^2 with coordinates x and y and consider the infinitesimal transformation $Z = x\dfrac{\partial}{\partial y} - y\dfrac{\partial}{\partial x}$. The 'infinitesimal transformation' defined by Z is given by the vector $(-y\,dt, x\,dt)$ at the point (x, y); this is a small vector at right angles to the radius vector and of length proportional to its distance from the origin O. In other words, Z represents an infinitesimal rotation about O and Zf is the rate of change of f under rotation about O. To make this clear we introduce polar coordinates r, θ. These form a chart on the region obtained by omitting the origin. We have $Zr = 0$, $Z\theta = 1$, and hence, by (21),

$$Z = Zr\frac{\partial}{\partial r} + Z\theta\frac{\partial}{\partial \theta} = \frac{\partial}{\partial \theta}.$$

Thus $Z = \dfrac{\partial}{\partial \theta}$, as we had been led to expect from geometrical considerations.

Ex. 1. If X and ω are given locally by (19) and (20), then

$$\langle X, \omega \rangle = \xi^i w_i.$$

Ex. 2. If f is an analytic function defined on the whole of $\mathfrak{M}$, then $df \in \mathfrak{L}^*$.

Ex. 3. If for each point $p \in \mathfrak{M}$ the space $\mathfrak{L}_p$ is spanned by the tangent vectors X_p $(X \in \mathfrak{L})$, show that a differential form ω is analytic if and only if $\langle X, \omega \rangle \in \mathscr{A}$ for all $X \in \mathfrak{L}$.

1.8. Mappings of manifolds. So far we have associated various sets such as $\mathscr{A}$, $\mathfrak{L}_p$, $\mathfrak{L}_p^*$, $\mathfrak{L}$, $\mathfrak{L}^*$ with a given analytic manifold $\mathfrak{M}$. When we deal with several analytic manifolds $\mathfrak{M}, \mathfrak{N}, \mathfrak{P}, \ldots$, we shall indicate in brackets to which manifold we are referring. Thus $\mathscr{A}(\mathfrak{M})$ will denote the set of analytic functions in $\mathfrak{M}$, $\mathfrak{L}_q(\mathfrak{N})$ the set of tangent vectors at $q \in \mathfrak{N}$, etc.

Let M and N be any manifolds, and consider the mapping $\Phi: p \to p^\Phi$ of M into N. For each function f defined in N we can define a function f^* in M by the rule

$$f^*(p) = f(p^\Phi) \quad (p \in M).$$

The function f^* so defined will be denoted by $\{\Phi\}f$; if D is the domain of $f(D \subseteq N)$, then the domain of $\{\Phi\}f$ is the inverse image of D under the mapping Φ. Suppose now that $\mathfrak{M}$ and $\mathfrak{N}$ are analytic manifolds. Then the mapping Φ of $\mathfrak{M}$ into $\mathfrak{N}$ is said to be *analytic*, if $\{\Phi\}f \in \mathscr{A}(\mathfrak{M})$ whenever $f \in \mathscr{A}(\mathfrak{N})$. Thus every analytic mapping Φ of $\mathfrak{M}$ into $\mathfrak{N}$ induces a mapping $\{\Phi\}$ of $\mathscr{A}(\mathfrak{N})$ into $\mathscr{A}(\mathfrak{M})$. In order to obtain an explicit expression for Φ we take a chart $(x) = (x^1, \ldots, x^m)$ in $\mathfrak{M}$; then the point p^Φ of $\mathfrak{N}$ depends on the coordinates x^i of p. By choosing a chart $(y) = (y^1, \ldots, y^n)$ on a suitable region of $\mathfrak{N}$ we may express this dependence as

$$y^i = \phi^i(x^1, \ldots, x^m) \quad (i = 1, \ldots, n). \tag{23}$$

If the mapping Φ is analytic, then the functions ϕ^i occurring in (23) will be analytic functions of their arguments. Conversely, if at each point p of $\mathfrak{M}$, Φ can be expressed by means of equations of the form (23), with analytic functions ϕ^i, then Φ is an analytic mapping, as is easily verified.

If I is the identity mapping of $\mathfrak{M}$ onto itself, defined by $p^I = p$ $(p \in \mathfrak{M})$, then I is analytic and

$$\{I\}f = f \quad (f \in \mathscr{A}(\mathfrak{M})). \tag{24}$$

Further, if $\mathfrak{M}$, $\mathfrak{N}$ and $\mathfrak{P}$ are analytic manifolds, and Φ, Ψ are analytic mappings of $\mathfrak{M}$ into $\mathfrak{N}$ and $\mathfrak{N}$ into $\mathfrak{P}$ respectively, then the mapping $\Phi\Psi$ of $\mathfrak{M}$ into $\mathfrak{P}$, i.e. the mapping obtained by performing Φ and Ψ in succession, is again analytic, and

$$\{\Phi\Psi\}f = \{\Phi\}\{\Psi\}f \quad (f \in \mathscr{A}(\mathfrak{P})). \tag{25}$$

These properties are an immediate consequence of the definitions and their verification is left to the reader. The properties stated in (24) and (25) are expressed by saying that the correspondence $\mathscr{F}$ which associates with the category of analytic manifolds $\mathfrak{M}$ and analytic mappings $\Phi: \mathfrak{M} \to \mathfrak{N}$, the category of sets of analytic functions and mappings by the rule $\mathscr{F}(\mathfrak{M}) = \mathscr{A}(\mathfrak{M})$, $\mathscr{F}(\Phi) = \{\Phi\}$, is a *contravariant functor*. The force of the prefix 'contra' is that $\{\Phi\}$ acts in a sense opposite to that of Φ. If Φ: $\mathfrak{M} \to \mathfrak{N}$, then $\mathscr{F}(\Phi)$: $\mathscr{F}(\mathfrak{N}) \to \mathscr{F}(\mathfrak{M})$. It is for this reason that we write Φ on the right and $\{\Phi\}$ on the left.

Let us suppose now that Φ is an analytic mapping of $\mathfrak{M}$ onto $\mathfrak{N}$ with an analytic inverse, i.e. there is an analytic mapping Φ^{-1} of $\mathfrak{N}$ into $\mathfrak{M}$ such that $\Phi\Phi^{-1}$ is the identity on $\mathfrak{M}$ and $\Phi^{-1}\Phi$ is the identity on $\mathfrak{N}$. Such a mapping Φ will be called an *analytic homeomorphism* of M onto N.† For an analytic homeomorphism we have, by (24) and (25),

$$\{\Phi\}\{\Phi^{-1}\}f=f \quad (f\in\mathscr{A}(\mathfrak{M})),$$
$$\{\Phi^{-1}\}\{\Phi\}g=g \quad (g\in\mathscr{A}(\mathfrak{N})).$$

Thus Φ induces a one-one mapping $\{\Phi\}$ of $\mathscr{A}(\mathfrak{N})$ onto $\mathscr{A}(\mathfrak{M})$, and $\{\Phi\}^{-1}=\{\Phi^{-1}\}$. Since $\{\Phi\}$ now has an inverse, we can calculate the transform of any mapping of $\mathscr{A}(\mathfrak{M})$ into itself: If X is a mapping of $\mathscr{A}(\mathfrak{M})$ into itself, then

$$g\to\{\Phi^{-1}\}X\{\Phi\}g \quad (g\in\mathscr{A}(\mathfrak{N})),$$

is a mapping of $\mathscr{A}(\mathfrak{N})$ into itself. We shall write $X^{d\Phi}$ instead of $\{\Phi^{-1}\}X\{\Phi\}$. Thus $d\Phi$ associates with every mapping of $\mathscr{A}(\mathfrak{M})$ into itself a mapping of $\mathscr{A}(\mathfrak{N})$ into itself.

$$\begin{array}{ccc} \mathscr{A}(\mathfrak{M}) & \xrightarrow{X} & \mathscr{A}(\mathfrak{M}) \\ \{\Phi\}\uparrow & & \uparrow\{\Phi\} \\ \mathscr{A}(\mathfrak{N}) & \xrightarrow{X^{d\Phi}} & \mathscr{A}(\mathfrak{N}) \end{array}$$

In particular, if $X\in\mathfrak{L}(\mathfrak{M})$, then

$$X^{d\Phi}\in\mathfrak{L}(\mathfrak{N}).$$

To prove this it is enough to show that the three conditions of Theorem 1.7.1 are satisfied. Let $p\in\mathfrak{M}$ and write $q=p^{\Phi}$; then for any $g\in\mathscr{A}_q(\mathfrak{N})$,

$$(X^{d\Phi}g)_q=(\{\Phi^{-1}\}X\{\Phi\}g)_q=(X\{\Phi\}g)_p=X_p\{\Phi\}g;$$

hence $X^{d\Phi}$ satisfies (i). Clearly $X^{d\Phi}$ is linear over R, and to verify (iii) we take f, g in $\mathscr{A}_q(\mathfrak{N})$ and put $\{\Phi\}f=f^*$, $\{\Phi\}g=g^*$ for brevity. Then, since X satisfies (iii) of Theorem 1.7.1, we have

$$\begin{aligned} X^{d\Phi}(fg) &= \{\Phi^{-1}\}X\{\Phi\}(fg) \\ &= \{\Phi^{-1}\}X(f^*g^*) \\ &= \{\Phi^{-1}\}(Xf^*.g^*+f^*.Xg^*) \\ &= \{\Phi^{-1}\}Xf^*.g+f.\{\Phi^{-1}\}Xg^* \\ &= X^{d\Phi}(f).g+f.X^{d\Phi}(g), \end{aligned}$$

which shows that $X^{d\Phi}$ satisfies (iii) as well and so belongs to $\mathfrak{L}(\mathfrak{N})$.

† It is clear that Φ, regarded as a mapping between topological spaces, is in fact a homeomorphism.

Thus $d\Phi$ is a mapping of $\mathfrak{L}(\mathfrak{M})$ into $\mathfrak{L}(\mathfrak{N})$. It is again not hard to verify that if I denotes the identity mapping on $\mathfrak{M}$ then

$$X^{dI} = X, \tag{26}$$

and if Φ, Ψ are analytic homeomorphisms of $\mathfrak{M}$ onto $\mathfrak{N}$ and $\mathfrak{N}$ onto $\mathfrak{P}$ respectively, then

$$d(\Phi . \Psi) = d\Phi . d\Psi. \tag{27}$$

Equations (26) and (27) are expressed by saying that the correspondence $\mathscr{G}$ defined by $\mathscr{G}(\mathfrak{M}) = \mathfrak{L}(\mathfrak{M})$, $\mathscr{G}(\Phi) = d\Phi$ is a *covariant functor* on the category of analytic manifolds and analytic homeomorphisms. Here $d\Phi$ acts in the same sense as Φ; if $\Phi : \mathfrak{M} \to \mathfrak{N}$, then $\mathscr{G}(\Phi) : \mathscr{G}(\mathfrak{M}) \to \mathscr{G}(\mathfrak{N})$.

THEOREM 1.8.1. *Let $\mathfrak{M}$ and $\mathfrak{N}$ be analytic manifolds and Φ an analytic homeomorphism of $\mathfrak{M}$ onto $\mathfrak{N}$. Then $d\Phi$ is an isomorphism of $\mathfrak{L}(\mathfrak{M})$ onto $\mathfrak{L}(\mathfrak{N})$, qua vector space over R.*

Proof. We saw already that Φ maps $\mathfrak{L}(\mathfrak{M})$ into $\mathfrak{L}(\mathfrak{N})$; by (26) and (27) $(d\Phi)^{-1} = d(\Phi^{-1})$, so the theorem will follow if we can show that $d\Phi$ is linear. This is an immediate consequence of the definition: If f, $g \in \mathscr{A}(\mathfrak{N})$, then

$$\{\Phi\}(\alpha f + \beta g) = \alpha\{\Phi\} f + \beta\{\Phi\} g \quad (\alpha, \beta \in R);$$

similarly with $\{\Phi\}$ replaced by $\{\Phi^{-1}\}$, and therefore

$$\begin{aligned}(\alpha X + \beta Y)^{d\Phi}(f) &= \{\Phi^{-1}\}(\alpha X + \beta Y)\{\Phi\}(f) \\ &= (\alpha X^{d\Phi} + \beta Y^{d\Phi})(f).\end{aligned}$$

This completes the proof.

In terms of the charts (x) at p and (y) at $q = p^{\Phi}$ which were used in writing down (23), the mapping $d\Phi$ may be expressed as follows: Let $X = \xi^i . \partial/\partial x^i$, then

$$X^{d\Phi} = \xi^i(\psi(y)) \frac{\partial \phi^j}{\partial x^i} \frac{\partial}{\partial y^j}, \tag{28}$$

where $x^i = \psi^i(y^1, \ldots, y^n)$ are the equations defining the mapping Φ^{-1}.

Still assuming Φ to be invertible, let us compare the tangent vectors at corresponding points in $\mathfrak{M}$ and $\mathfrak{N}$. If $p \in \mathfrak{M}$ and $p^{\Phi} = q$ say, then for any $g \in \mathscr{A}_q(\mathfrak{N})$,

$$(X^{d\Phi} g)_q = (\{\Phi^{-1}\} X \{\Phi\} g)_q = (X\{\Phi\} g)_p = X_p \{\Phi\} g.$$

Since the result depends only on X_p and not on the whole of X, $d\Phi$ defines a mapping of tangent vectors at p into tangent vectors at $q = p^\Phi$. This mapping is clearly linear, and since it has an inverse, defined by $d\Phi^{-1}$, it establishes an isomorphism between $\mathfrak{L}_p(\mathfrak{M})$ and $\mathfrak{L}_q(\mathfrak{N})$. It follows that these spaces have the same dimension, and so we obtain

THEOREM 1.8.2. *If $\mathfrak{M}$ and $\mathfrak{N}$ are analytic manifolds and Φ is an analytic homeomorphism of $\mathfrak{M}$ onto $\mathfrak{N}$, then $\mathfrak{M}$ and $\mathfrak{N}$ have the same dimension at corresponding points.*

The theorem follows from the preceding remarks and Theorem 1.4.2.

We shall sometimes be concerned with local properties of mappings, and we now turn to consider the most important of these. A mapping Φ of $\mathfrak{M}$ into $\mathfrak{N}$ is said to be *analytic at a point p of $\mathfrak{M}$*, if $\{\Phi\}f \in \mathscr{A}_p(\mathfrak{M})$ for all $f \in \mathscr{A}_q(\mathfrak{N})$, where $q = p^\Phi$. We are particularly interested to know under what conditions the mapping Φ has a 'local inverse' at a point p_0 of $\mathfrak{M}$, i.e. when there is a mapping Ψ of a neighbourhood W of $q_0 = p_0^\Phi$ onto a neighbourhood V of p_0 such that

$$p^{\Phi\Psi} = p \quad (p \in V),$$
$$q^{\Psi\Phi} = q \quad (q \in W).$$

If Φ possesses a local inverse at p_0 which is analytic at p_0^Φ, we shall say that Φ is *invertible at p_0*.

THEOREM 1.8.3. *Let Φ be an analytic mapping of $\mathfrak{M}$ into $\mathfrak{N}$, and suppose that in terms of charts $(x^1, \ldots, x^m)$ at $p \in \mathfrak{M}$ and $(y^1, \ldots, y^n)$ at $q = p^\Phi \in \mathfrak{N}$, Φ takes the form*

$$y^i = \phi^i(x^1, \ldots, x^m).$$

Then Φ is invertible at p provided that $m = n$ and $\det(\partial\phi^i/\partial x^j)_p \neq 0$.

Proof. Suppose that the hypotheses are satisfied. Define n real functions $'x^1, \ldots, 'x^n$ in $\mathfrak{M}$ by

$$'x^i = \phi^i(x^1, \ldots, x^n),$$

then $d'x^i = [\partial\phi^i/\partial x^j]_p dx^j$ are n linearly independent differentials at p, since by hypothesis the matrix of coefficients is non-singular. Hence they form a basis of $\mathfrak{L}_p(\mathfrak{M})$, and by Theorem 1.6.2, $('x^1, \ldots, 'x^n)$ is an admissible chart at p. The equations of the mapping Φ, when expressed in terms of the new chart, take the form

$$y^i = 'x^i.$$

The mapping defined near q by the equations

$$'x^i = y^i$$

is clearly a local inverse of Φ, analytic at q, and the theorem follows.

From this proof we see that Φ need only be defined near p for the conclusion of the theorem to hold. The proof also shows that if an analytic mapping of $\mathfrak{M}$ into $\mathfrak{N}$ is invertible at p, then we may take charts at p and p^Φ such that points which correspond under the mapping Φ have the same coordinates. Thus Φ may be used to transfer charts from $\mathfrak{M}$ to $\mathfrak{N}$ and conversely, provided that we restrict ourselves to suitable neighbourhoods of p and p^Φ.

Ex. 1. Show that the conditions of Theorem 1.8.3 are necessary as well as sufficient.

Ex. 2. Show that the mapping $t \to t^3$ of $\mathfrak{R}$ into itself is analytic and has a local inverse at every point, but has no analytic local inverse at $t=0$.

Ex. 3. Each element A of $\mathfrak{L}(\mathfrak{M})$ defines an 'infinitesimal' mapping

$$\Phi\colon\ x^i \to x^i + Ax^i \,.\, \delta t$$

of $\mathfrak{M}$ into itself. Verify that this definition is independent of the chart used and show that, to the first order in δt, Φ has the inverse

$$\Phi^{-1}\colon\ x^i \to x^i - Ax^i \,.\, \delta t.$$

If $X \in \mathfrak{L}(\mathfrak{M})$, show that $X^{d\Phi}f = Xf + \{X(Af) - A(Xf)\} \,.\, \delta t$.

1.9. Submanifolds. In an n-dimensional Euclidean space a ν-dimensional subspace may be specified by $n-\nu$ equations between the n coordinates, or by expressing the n coordinates as functions of ν independent parameters; or as a combination of these two methods, by expressing $n-\nu$ suitably chosen coordinates in terms of the remaining ν. We shall use this last form to arrive at a suitable generalization for manifolds.

Suppose that we have a real function $f=f(p)$ defined in a manifold M. If we restrict the argument p to a subset W of M, we obtain a real function defined in W—the restriction of f to W—which we denote by $f \mid W$.

DEFINITION. Given two analytic manifolds $\mathfrak{M}$ and $\mathfrak{N}$, we say that $\mathfrak{N}$ is a *submanifold* of $\mathfrak{M}$, if $\mathfrak{N}$ is a subset of $\mathfrak{M}$ such that

SM. *For any admissible chart* $(x^1, \ldots, x^n)$ *in* $\mathfrak{M}$, *the functions* $x^i \mid \mathfrak{N}$ *are analytic functions in* $\mathfrak{N}$, *and at each point* p *of* $\mathfrak{N}$ *at which they are defined we can select a subset* $(x^{i_1} \mid \mathfrak{N}, \ldots, x^{i_\nu} \mid \mathfrak{N})$ *which forms an admissible chart at* p.

It is a consequence of SM that the identity mapping of $\mathfrak{N}$ into $\mathfrak{M}$ is analytic; in general this condition is not strong enough to replace SM, as Ex. 4 at the end of this section shows, but we shall see in Chapter VI that the condition *is* equivalent to SM in the case in which we are interested, namely, that of Lie groups.

As usual, it is enough to postulate SM for a family of charts covering $\mathfrak{M}$. For if at a given point p of $\mathfrak{N}$ the property SM holds for one chart (x), then it also holds for any other chart (y) at p. To prove this statement, let us denote $x^i \mid \mathfrak{N}$ by $\bar{x}^i$ for short; we may suppose, by renumbering the x^i if necessary, that the set $(\bar{x}^1, \ldots, \bar{x}^\nu)$ forms in $\mathfrak{N}$ an admissible chart at p. The charts (x) and (y) are analytically related:

$$x^i = \phi^i(y) \quad (\phi^i \text{ analytic, } i = 1, \ldots, n), \tag{29}$$

$$y^i = \psi^i(x) \quad (\psi^i \text{ analytic, } i = 1, \ldots, n). \tag{30}$$

Moreover, by the hypothesis for (x),

$$\bar{x}^A = \theta^A(\bar{x}^1, \ldots, \bar{x}^\nu) \quad (\theta^A \text{ analytic, } A = \nu + 1, \ldots, n). \tag{31}$$

When we restrict the arguments in (29) and (30) to $\mathfrak{N}$ we obtain

$$\bar{x}^i = \phi^i(\bar{y}^1, \ldots, \bar{y}^n), \tag{32}$$

$$\bar{y}^i = \psi^i(\bar{x}^1, \ldots, \bar{x}^n). \tag{33}$$

Inserting (31) in (33), we obtain

$$\bar{y}^i = \chi^i(\bar{x}^1, \ldots, \bar{x}^\nu), \tag{34}$$

where the functions χ^i are analytic. It follows that $\bar{y}^i \in \mathscr{A}_p(\mathfrak{N})$, and if we differentiate (32), we have at p,

$$d\bar{x}^i = \alpha^i_j d\bar{y}^j, \tag{35}$$

where the α^i_j are certain constants. Since $(\bar{x}^1, \ldots, \bar{x}^\nu)$ is a chart, the elements $d\bar{x}^1, \ldots, d\bar{x}^\nu$ form a basis of $\mathfrak{L}^*(\mathfrak{N})$; by (35) the elements $d\bar{y}^1, \ldots, d\bar{y}^n$ span $\mathfrak{L}^*(\mathfrak{N})$, and we can therefore choose a subset which forms a basis. This basis has again ν elements and by suitable numbering of the y's, may be taken to be $d\bar{y}^1, \ldots, d\bar{y}^\nu$. By Theorem 1.6.2, $(\bar{y}^1, \ldots, \bar{y}^\nu)$ is then an admissible chart at p, which proves that the chart (y) satisfies SM.

Let $\mathfrak{M}$ be an analytic manifold and $\mathfrak{N}$ a submanifold. At a point p of $\mathfrak{N}$, let $(x^1, \ldots, x^n)$ and $(y^1, \ldots, y^\nu)$ be charts, taken in $\mathfrak{M}$ and $\mathfrak{N}$ respectively. If we denote the restriction of x^i to $\mathfrak{N}$ again by $\bar{x}^i$, then the $\bar{x}^i$ are analytic functions of the y's:

$$\bar{x}^i = \chi^i(y^1, \ldots, y^\nu), \tag{36}$$

and these equations can be solved for $y^1, \ldots, y^\nu$ in terms of ν of the $\bar{x}^i$. It follows by Theorem A 3 (Appendix) that the matrix $(\partial\chi^i/\partial y^\rho)_p$ has rank ν. Conversely, if $\mathfrak{N}$ is an analytic manifold which is contained in another analytic manifold $\mathfrak{M}$, and if at any point of $\mathfrak{N}$ two charts in $\mathfrak{M}$ and $\mathfrak{N}$ respectively are related by equations (36) with the Jacobian matrix $(\partial\chi^i/\partial y^\rho)_p$ of rank ν, then $\mathfrak{N}$ is a submanifold of $\mathfrak{M}$. For then we can solve (36) in terms of ν of the $\bar{x}^i$, and these $\bar{x}$'s therefore form an admissible chart in $\mathfrak{N}$. This proves

THEOREM 1.9.1. *Let $\mathfrak{M}$ and $\mathfrak{N}$ be analytic manifolds such that $\mathfrak{N} \subseteq \mathfrak{M}$. Then $\mathfrak{N}$ is a submanifold of $\mathfrak{M}$ if and only if, at each point p of $\mathfrak{N}$, there are admissible charts (x) and (y) in $\mathfrak{M}$ and $\mathfrak{N}$ respectively, which are related by equations*

$$\bar{x}^i = \chi^i(y),$$

where the χ^i are analytic and the rank of $(\partial\chi^i/\partial y^\rho)_p$ is equal to the dimension of $\mathfrak{N}$ at p.

In an important special case, submanifolds can be constructed by

THEOREM 1.9.2. *Let $\mathfrak{M}$ be an analytic manifold and N a subset of $\mathfrak{M}$ which is open in the topology of $\mathfrak{M}$. Then there exists an analytic structure on N and the resulting analytic manifold $\mathfrak{N}$ has the properties:*

(i) *$\mathfrak{N}$ is a submanifold of $\mathfrak{M}$, of the same dimension as $\mathfrak{M}$ at each of its points,*

(ii) *the topology induced on N by the analytic structure of $\mathfrak{M}$ coincides with that induced by the analytic structure of $\mathfrak{N}$.*

Proof. Since N is open, it is a neighbourhood of each of its points, and for each $p \in N$ we can select a chart from the analytic structure of $\mathfrak{M}$ which lies in N. In this way we obtain an analytic family of charts covering N, and by Theorem 1.2.1 this defines an analytic structure on N. If we denote the resulting analytic manifold by $\mathfrak{N}$, then the topologies induced on N by $\mathfrak{M}$ and $\mathfrak{N}$ respectively coincide, and $\mathfrak{N}$ is a submanifold of $\mathfrak{M}$ because the chart selected at p in $\mathfrak{M}$ clearly satisfies SM. Thus the proof is complete.

We consider a few examples to illustrate the notion of a submanifold.

1. Let $\mathfrak{M} = \mathfrak{R}^3$ and $\mathfrak{N} = \mathfrak{R}^2$; $\mathfrak{N}$ is the (x, y)-plane in $\mathfrak{M}$, say. Further, let $\mathfrak{P}$ be the interior of the unit circle in the (x, y)-plane; $\mathfrak{P}$ is open in $\mathfrak{N}$ and therefore, by Theorem 1.9.2, it may be regarded as a submanifold of $\mathfrak{N}$. It is also clear that $\mathfrak{N}$ is a submanifold of $\mathfrak{M}$. The dimensions of $\mathfrak{M}$, $\mathfrak{N}$ and $\mathfrak{P}$ are 3, 2, 2 respectively. Thus the dimension of a submanifold of $\mathfrak{M}$ at a given point p may be less than or equal to the dimension of $\mathfrak{M}$ itself at p. Clearly it cannot be greater.

2. Let $\mathfrak{M} = \mathfrak{T}^2$; the coordinates may be taken as (x, y), where x and y are real numbers mod 1. For any real α, the set of points given by $x = t$, $y = \alpha t \pmod 1$, where t is a real parameter, is a 1-dimensional submanifold $\mathfrak{N}$ of $\mathfrak{M}$. If α is rational, $\mathfrak{N}$ is a closed curve on $\mathfrak{M}$, but for irrational α, $\mathfrak{N}$ is everywhere dense in $\mathfrak{M}$ and its topology is different from the one induced by $\mathfrak{M}$. To prove the last assertion, let p_n, q_n be integers such that $p_n - \alpha q_n \to 0$, then the points $(0, \alpha q_n)$ are in $\mathfrak{N}$ and they come arbitrarily close to $(0, 0)$ in $\mathfrak{M}$ but not in $\mathfrak{N}$. The rest is obvious.

3. Let us take the analytic manifold $\mathfrak{R}^2$ with the coordinates x and y, say. Secondly, we define an analytic structure on R^2 by taking as coordinate neighbourhood of any point the line through this point parallel to the x-axis, and the x-coordinate on this line as coordinate. With this definition the set R^2 becomes a 1-dimensional analytic manifold (the effect is just that of cutting up the plane into parallel lines); denote this analytic manifold by $\mathfrak{S}$. Then $\mathfrak{S}$ is a submanifold of $\mathfrak{R}^2$, but $\mathfrak{R}^2$ itself is also a submanifold of $\mathfrak{R}^2$. Thus a given subset of an analytic manifold may be a submanifold in several different ways.

Ex. 1. Any analytic manifold $\mathfrak{M}$ is a submanifold of itself.

Ex. 2. A submanifold of a submanifold of $\mathfrak{M}$ is again a submanifold of $\mathfrak{M}$.

Ex. 3. A submanifold of $\mathfrak{M}$ which has the same dimension as $\mathfrak{M}$ at each of its points is an open subset of $\mathfrak{M}$. (This may be regarded as the converse of Theorem 1.9.2.)

Ex. 4. Show that the plane curve $x=t^2$, $y=t^3$ possesses an analytic structure for which (t) is an admissible chart. If $\mathfrak{C}$ is the resulting analytic manifold then the identity mapping of $\mathfrak{C}$ into the (x, y)-plane $\mathfrak{R}^2$ is analytic, but $\mathfrak{C}$ is not a submanifold of $\mathfrak{R}^2$.

1.10. Products of manifolds. Let M and N be two manifolds; regarding M and N as topological spaces, we can form their topological product. This is by definition the set $M \times N$ of pairs (p, q) $(p \in M,\ q \in N)$ with the topology obtained by taking as a basis of open sets the sets $U \times V$, where U and V run over the open sets of M and N respectively. The space $M \times N$ is again a manifold, and if we start with two *analytic* manifolds $\mathfrak{M}$ and $\mathfrak{N}$, their product can be defined as an analytic manifold in a natural way. We merely give the proof in the analytic case.

In order to define the analytic structure on $\mathfrak{M} \times \mathfrak{N}$, it is enough to specify an analytic family of charts covering $M \times N$. Let (p_0, q_0), where $p_0 \in \mathfrak{M}$, $q_0 \in \mathfrak{N}$, be any element of $\mathfrak{M} \times \mathfrak{N}$. Then there is a chart $(x^1, \ldots, x^n)$ defined on a neighbourhood U of p_0, and a chart $(y^1, \ldots, y^m)$ defined on a neighbourhood V of q_0. More precisely, the coordinates of any point p of U are $x^i(p)$ and the coordinates of a point q of V are $y^j(q)$. Now $U \times V$ is a typical neighbourhood of the point (p_0, q_0) of $\mathfrak{M} \times \mathfrak{N}$ and the correspondence

$$(p, q) \leftrightarrow (x^1(p), \ldots, x^n(p),\ y^1(q), \ldots, y^m(q))$$

is a one-one correspondence between $U \times V$ and an open subset of R^{n+m}. This correspondence is easily verified to be a homeomorphism, by the definition of the product topology on $M \times N$. Hence $(x^1, \ldots, x^n,\ y^1, \ldots, y^m)$, or (x, y) for short, is a chart at (p_0, q_0). In this way, starting with two families of admissible charts covering $\mathfrak{M}$ and $\mathfrak{N}$ respectively, we obtain a family of charts covering $\mathfrak{M} \times \mathfrak{N}$. If (x, y) and (x^*, y^*) are any members of this family, then (x) and (x^*) are analytically related, and so are

(y) and (y^*). Since (x) depends on (x^*) alone and not on (y^*), while (y) depends only on (y^*) and not on (x^*) (and similarly with starred and unstarred coordinates interchanged), it follows that (x, y) and (x^*, y^*) are analytically related. Thus we have an analytic family of charts covering $\mathfrak{M} \times \mathfrak{N}$. By Theorem 1.2.1, this may be used to define $\mathfrak{M} \times \mathfrak{N}$ as an analytic manifold. We denote this analytic manifold again by $\mathfrak{M} \times \mathfrak{N}$ and call it the *product manifold* of $\mathfrak{M}$ and $\mathfrak{N}$.

It is clear how this definition may be extended to a product of more than two factors, but this will not be required for our purpose.

Ex. If $\mathfrak{M}$ and $\mathfrak{N}$ are of dimensions m and n respectively, then $\mathfrak{M} \times \mathfrak{N}$ is of dimension $m+n$.

CHAPTER II

TOPOLOGICAL GROUPS AND LIE GROUPS

2.1. Topological groups.

DEFINITION. A *topological group* is a set G with the following properties:

TG. 1. *G is a group, i.e. there is a multiplication defined on G which satisfies the group axioms,*

TG. 2. *G is a Hausdorff space,*

TG. 3. *The mapping $(x, y) \to xy^{-1}$ of $G \times G$ into G is continuous.*

Thus the set G has two structures defined on it, one algebraic and one topological, and they are connected by TG. 3. We express TG. 3 by saying that the topology on G is *compatible* with the group structure.

Illustrations: 1. Discrete groups. The discrete topology on an abstract group G is always compatible with the group structure. For if G has the discrete topology, then so has $G \times G$ and any mapping of a discrete space is continuous.

2. Let R be the additive group of real numbers. The topology defined by the metric $|a-b|$ is compatible with the group structure on R. When we speak of R as a topological group it is always this topology that we have in mind.

3. Let P be the additive group of rational numbers; P may be regarded as a subgroup of R and the topology induced on P by the topology of R (cf. 2) is compatible with the group structure.

4. Let R be the same group as in 2 but with the topology which is obtained by taking the half-intervals

$$H_\delta(a) = \{x \in R \mid a \leqslant x < a + \delta\} \quad (\delta > 0)$$

to be a neighbourhood base at a point $a \in R$. In this topology addition is a continuous operation, but not subtraction. So this does not define a topological group.

We shall generally use the multiplicative notation for groups and denote the unit element, or identity, of any group by e, when this does not lead to confusion. If S and T are any subsets of a group, we denote by ST the set of all elements st, where $s \in S$, $t \in T$, and by S^{-1} the set of all s^{-1} $(s \in S)$. When S consists of a single element s, we shall also write sT instead of ST, and similarly for T.

The compatibility condition TG. 3 is equivalent to the following two conditions:

TG. 3′. *The mapping* $(x, y) \to xy$ *of* $G \times G$ *into* G *is continuous*,

TG. 3″. *The mapping* $x \to x^{-1}$ *of* G *into* G *is continuous.*

For if TG. 3 holds, then putting $x = e$, we see that y^{-1} is continuous in y, and hence $xy = x(y^{-1})^{-1}$ is continuous in x and y. Conversely, if TG. 3′ and TG. 3″ are satisfied, then $(x, y) \to (x, y^{-1})$ is a continuous mapping of $G \times G$ into itself, and hence the mapping $(x, y) \to (x, y^{-1}) \to xy^{-1}$ is continuous, so that TG. 3 holds.

In terms of neighbourhoods, TG. 3 states that for any elements x and y of a topological group G, and any neighbourhood W of xy^{-1}, there exist neighbourhoods U, V of x and y respectively, such that $UV^{-1} \subseteq W$. Correspondingly, TG. 3′ states that for any neighbourhood W of xy there exist neighbourhoods U, V of x and y respectively, such that $UV \subseteq W$, while TG. 3″ states that for any neighbourhood U of x, U^{-1} is a neighbourhood of x^{-1}.

Now let G be any topological group. The mapping $x \to x^{-1}$ is a continuous mapping of G into itself; since $(x^{-1})^{-1} = x$, the mapping coincides with its inverse. Thus the inverse exists and is also continuous, whence the mapping $x \to x^{-1}$ is a homeomorphism of G.

Any element a of G defines a mapping

$$\rho_a \colon x \to xa \quad (x \in G)$$

of G into itself, which is called a *right translation*. This mapping ρ_a is in fact a one-one mapping of G onto itself, since it possesses the inverse $\rho_{a^{-1}}$. The mapping ρ_a is continuous, since xa is a continuous function of x; for a like reason $\rho_{a^{-1}}$ is continuous, and so ρ_a is a homeomorphism of G. The *left translation* associated with a is defined similarly as

$$\lambda_a \colon x \to ax \quad (x \in G),$$

and this is also seen to be a homeomorphism of G.

The importance of translations is that they provide homeomorphisms of G with itself which map any given point into any other given point. Thus, given any two points a, $b \in G$, we can find a right translation, namely, $\rho_{a^{-1}b}$, which maps a into b. The left translation $\lambda_{ba^{-1}}$ also has this property. This proves

THEOREM 2.1.1. *Given any two points a, b of a topological group G, there exists a homeomorphism of G which maps a into b.*

A Hausdorff space with the property mentioned in this theorem is called a *homogeneous space.*

By Theorem 2.1.1, if a topological group has some local property at a given point, say at e, then it has it everywhere. Thus it is sufficient to verify such properties at e. The next theorem shows that this applies to the continuity of the group operations themselves.

THEOREM 2.1.2. *Let G be a group which is also a topological space. Then the topology on G is compatible with the group operations if and only if,*

(i) *the translations ρ_a, λ_b $(a, b \in G)$ are continuous,*

(ii) *the mapping $(x, y) \to xy^{-1}$ is continuous at the point (e, e) of $G \times G$.*

Proof. The necessity of the conditions is clear from the definitions and the remarks preceding Theorem 2.1.1. Suppose now that (i) and (ii) hold. We have to show that the mapping $(x, y) \to xy^{-1}$ is continuous at the general point (a, b) of $G \times G$. If we put $x = au$, $y = bv$, then by (ii), $(u, v) \to uv^{-1}$ is continuous at (e, e), and using (i), we find that at (a, b), the mapping

$$(x, y) \to (a^{-1}x, b^{-1}y) = (u, v) \to uv^{-1} \to auv^{-1}b^{-1} = xy^{-1}$$

is continuous, as we wished to prove.

In this theorem we only assumed that G has a topology which is compatible with the group operations. If we want to find out whether G is a topological group we have to see whether TG. 2 holds. This axiom can still be weakened without affecting the definition of a topological group. In fact, it is enough to assume that G is a T_1-space (Fréchet space). Since this is a local property, we need only assume that it is satisfied at e:

THEOREM 2.1.3. *Let G be a group and suppose that a topology is defined on G which is compatible with the group operations. If, further, $\{e\}$ is closed, then G is a topological group.*

For let $a, b \in G$ be such that $a \neq b$; since $\{e\}$ is closed, it follows that $\{a\} = \{e\}\rho_a$ is closed, and so there is a neighbourhood V of b such that $V \cap \{a\} = \emptyset$.† The set V has the form Wb, where W is a neighbourhood of e. Let U be a neighbourhood of e such that $U^{-1}U \subseteq W$. We shall prove that G is a Hausdorff space by showing that

$$Ua \cap Ub = \emptyset. \tag{1}$$

If $x \in Ua \cap Ub$, say, then $x = u_1 a = u_2 b$ $(u_1, u_2 \in U)$, and so

$$a = u_1^{-1} u_2 b \in U^{-1}Ub \subseteq Wb = V,$$

which contradicts the fact that $V \cap \{a\} = \emptyset$. Hence (1) holds and the theorem follows.

Ex. 1. If τ denotes the mapping $x \to x^{-1}$, show that $\lambda_a = \tau\rho_{a^{-1}}\tau$.

Ex. 2. Theorem 2.1.2 remains valid if in (i) only the continuity of τ, defined as in Ex. 1, and all the right translations (or of τ and all the left translations) is assumed.

Ex. 3. Show that any connected manifold is a homogeneous space. (Prove this first for a sphere in R^n and then apply the method of Theorem 7.1.1.)

2.2. The family of nuclei of a topological group.

In any topological space the topology is completely defined by the neighbourhoods of its points, and since in a topological group these neighbourhoods may all be obtained by translation from the neighbourhoods of e, it should be possible to define a topological group in terms of its neighbourhoods of e. This is done in Theorem 2.2.1, which gives necessary and sufficient conditions for a family of subsets of a group G to form the set of neighbourhoods of e in some topology compatible with the group structure. For brevity, a neighbourhood of the unit element of a topological group G will in future be called a *nucleus* of G.

THEOREM 2.2.1. *If G is a topological group and $\mathscr{V}$ the family of all nuclei of G, then*

V.1. *If $V_1, V_2 \in \mathscr{V}$, then $V_1 \cap V_2 \in \mathscr{V}$,*

V.2. *If $V \in \mathscr{V}$ and $W \supseteq V$, then $W \in \mathscr{V}$,*

† $\emptyset$ denotes the empty set.

V.3. $\bigcap_{V \in \mathscr{V}} V = \{e\}$,

V.4. *Given* $V \in \mathscr{V}$, *there exists* $V_1 \in \mathscr{V}$ *such that* $V_1 V_1^{-1} \subseteq V$,

V.5. *If* $V \in \mathscr{V}$ *and* $a \in G$, *then* $a^{-1} V a \in \mathscr{V}$.

Conversely, given a group G and a family $\mathscr{V}$ *of subsets of G satisfying* V.1–5, *then there is a uniquely determined Hausdorff topology on G which is compatible with the group structure, and such that* $\mathscr{V}$ *is the family of nuclei in this topology.*

Proof. If $\mathscr{V}$ is the family of nuclei in a topological group, then V.1, 2 follow from the definition of neighbourhood and V.3 follows because each point is closed: given $a \neq e$, there is a nucleus not containing a. V.4 expresses the continuity of the mapping $(x, y) \rightarrow xy^{-1}$ at (e, e), while V.5 expresses the continuity of the mapping $x \rightarrow a^{-1}xa$ at e.

Now suppose conversely that $\mathscr{V}$ is a family of subsets of a group G satisfying V.1–5. We have to define a topology on G with the requisite properties, and we may do this, for example, in terms of open sets. So let $\mathscr{O}$ be the family of subsets O of G such that

$$x \in O \text{ implies } Vx \subseteq O \text{ for some } V \in \mathscr{V}. \tag{2}$$

The rest of the proof proceeds in a number of steps.

1. $\mathscr{O}$ satisfies the axioms for a family of open sets. Clearly $\emptyset, G \in \mathscr{O}$, and if $O_\alpha \in \mathscr{O} (\alpha \in A)$, then $\bigcup_{\alpha \in A} O_\alpha \in \mathscr{O}$. Further, if $O_1, O_2 \in \mathscr{O}$ and x is any element of $O_1 \cap O_2$, then by (2) there exist $V_1, V_2 \in \mathscr{V}$ such that $V_1 x \subseteq O_1$, $V_2 x \subseteq O_2$, hence $(V_1 \cap V_2) x \subseteq O_1 \cap O_2$. But $V_1 \cap V_2 \in \mathscr{V}$ by V.1, hence $O_1 \cap O_2 \in \mathscr{O}$. Thus $\mathscr{O}$ may be taken to be the family of open sets of a topology on G, and this topology is completely defined by $\mathscr{O}$.

2. Let $\mathscr{W}$ be the family of neighbourhoods of e in this topology; we show next that $\mathscr{W} = \mathscr{V}$. Suppose that $W \in \mathscr{W}$; then there exists $O \in \mathscr{O}$ such that $e \in O \subseteq W$, and hence there is a $V \in \mathscr{V}$ such that $V = Ve \subseteq O \subseteq W$. Therefore $W \in \mathscr{V}$ by V.2, and so $\mathscr{W} \subseteq \mathscr{V}$. To prove the reverse inclusion, let $V \in \mathscr{V}$ and define U as the set of all $x \in V$ such that† $V_1 x \subseteq V$ for some $V_1 \in \mathscr{V}$. Clearly $U \subseteq V$, and $e \in U$, since $Ve \subseteq V$. We shall show that $U \in \mathscr{O}$. For this purpose we first remark that

$$\text{Given } V \in \mathscr{V}, \text{ there exists } S \in \mathscr{V} \text{ such that } SS \subseteq V. \tag{3}$$

† U will turn out to be the interior of V.

For by V.4 there exists $V_1 \in \mathscr{V}$ such that $V_1 V_1^{-1} \subseteq V$. Hence $V_1^{-1} = eV_1^{-1} \subseteq V_1 V_1^{-1} \subseteq V$, and so $V_1 \subseteq V^{-1}$, which proves by V.2 that $V^{-1} \in \mathscr{V}$. This holds for any $V \in \mathscr{V}$, and in particular for V_1 itself. Let $S = V_1 \cap V_1^{-1}$, then $S \in \mathscr{V}$ by V.1, and $SS \subseteq V_1 V_1^{-1} \subseteq V$, which establishes (3).

Now let $a \in U$, then by definition there exists $V_1 \in \mathscr{V}$ such that $V_1 a \subseteq V$. By (3) $V_2 V_2 \subseteq V_1$ for some $V_2 \in \mathscr{V}$, hence $V_2 V_2 a \subseteq V_1 a \subseteq V$, i.e. for any $b \in V_2 a$ we have $V_2 b \subseteq V$. Therefore $V_2 a \subseteq U$, and since a was any element of U, this proves that $U \in \mathscr{O}$. Thus U is an open subset of V containing e, whence V is a neighbourhood of e, i.e. $V \in \mathscr{W}$. This shows that $\mathscr{V} \subseteq \mathscr{W}$ and so $\mathscr{W} = \mathscr{V}$.

3. The topology is compatible with the group structure. It is clear from (2) that for any $a \in G$, O is open whenever Oa is open, and thus the right translation ρ_a is continuous. To show that λ_a is continuous, let $Q \subseteq G$ and suppose that $aQ \in \mathscr{O}$. If $x \in Q$ then $ax \in aQ$, whence $Vax \subseteq aQ$ for some $V \in \mathscr{V}$, and so $a^{-1}Vax \subseteq Q$. But $a^{-1}Va \in \mathscr{V}$ by V.5 and so $Q \in \mathscr{O}$, which proves the continuity of λ_a. Finally, V.4 just expresses that the mapping $(x, y) \to xy^{-1}$ is continuous at (e, e), and applying Theorem 2.1.2, we see that the topology is compatible with the group structure.

4. To prove the Hausdorff property we need only verify, by Theorem 2.1.3, that $\{e\}$ is closed. Let $a \neq e$, then $a^{-1} \neq e$, and by V.3 there is a $V \in \mathscr{V}$ not containing a^{-1}, therefore $e \notin Va$, and so $a \notin \overline{\{e\}}$. This shows that $\overline{\{e\}}$ consists of e alone.

5. The uniqueness of the topology follows from the fact that it is completely determined by $\mathscr{V}$ as the family of nuclei of G. This completes the proof of Theorem 2.2.1.

2.3. Subgroups and homomorphic images.

If G is a topological group, a *subgroup* of G is a subset H such that $HH^{-1} \subseteq H$, i.e. a subgroup in the abstract sense. We shall often deal with subgroups which, qua topological spaces, are closed; they will be called *closed subgroups*.†

Let G be a topological group and H a subgroup, not necessarily closed. We denote the set of right cosets Ha of H in G by G/H, and we call the mapping

$$\phi: a \to Ha$$

† The adjective is absorbed into the definition of 'subgroup' by some writers.

of G onto G/H the *natural mapping* associated with H. The topology of G induces a topology on H, and it is not hard to verify that this topology is compatible with the group structure of H, so that we may regard H again as a topological group. We shall now show that the topology of G also induces a topology on G/H, in a sense which is made precise by

THEOREM 2.3.1. *If H is any subgroup of a topological group G, then G/H can be defined as a topological space in such a way that*

(i) *the natural mapping ϕ of G onto G/H is continuous,*

(ii) *if ψ is any mapping of G/H into a topological space T such that the mapping $\phi\psi$ of G into T is continuous, then ψ is continuous.*† *The topology thus defined on G/H is uniquely determined by* (i) *and* (ii).

This topology will be called the topology *induced* on G/H by G.

Proof. We define a topology on G/H as follows: a subset K of G/H is open in G/H if and only if its inverse image $K\overset{-1}{\phi}$ is open in G. From the rules

$$\bigcup_\alpha (K_\alpha \overset{-1}{\phi}) = (\bigcup_\alpha K_\alpha)\overset{-1}{\phi},$$

$$\bigcap_\alpha (K_\alpha \overset{-1}{\phi}) = (\bigcap_\alpha K_\alpha)\overset{-1}{\phi}$$

(which are easily verified), it follows that this definition satisfies the axioms for open sets. By definition, if K is open in G/H, then $K\overset{-1}{\phi}$ is open in G, hence ϕ is continuous. To prove (ii), let ψ be a mapping of G/H into a space T such that $\phi\psi$ is continuous. Given any open subset O of T, it follows that $(O\overset{-1}{\psi})\overset{-1}{\phi}$ is open in G; hence $O\overset{-1}{\psi}$ is open in G/H, and this proves the continuity of ψ. Thus there exists a topology $\mathscr{T}$ on G/H satisfying (i) and (ii). Now if we are given any topology $\mathscr{T}'$ on G/H which satisfies (i) and (ii), then the identity mapping ψ of G/H with the topology $\mathscr{T}$ onto G/H with the topology $\mathscr{T}'$ has the property that $\phi\psi$ is continuous, because $\mathscr{T}'$ satisfies (i). Hence ψ must be continuous, since $\mathscr{T}$ satisfies (ii). Reversing the roles of $\mathscr{T}$ and $\mathscr{T}'$, we find that $\overset{-1}{\psi}$ is also continuous, and therefore ψ is a homeomorphism. Thus $\mathscr{T} = \mathscr{T}'$ and the theorem is completely proved.

† These mappings are written on the right. Thus $\phi\psi$ is the mapping obtained by performing first ϕ, then ψ.

Theorem 2.3.1 expresses the fact that the topology induced by G on G/H is the finest topology for which the natural mapping ϕ of G onto G/H is continuous. We observe further that ϕ maps open sets into open sets; if K is an open subset of G, then aK is open for all $a \in G$ and hence $HK = \bigcup_{h \in H} hK$ is open, but

$$HK = (K\phi)\overset{-1}{\phi};$$

therefore $K\phi$ is open. We express this fact by saying that ϕ is an *open* mapping.

THEOREM 2.3.2. *Let N be a normal subgroup of a topological group G, and let G/N be the quotient group. Then the topology induced on G/N by G is compatible with the group structure of G/N, and if, further, N is closed, then G/N becomes a topological group in this way. The natural mapping ϕ of G onto G/N is then an open continuous homomorphism.*

We have to show that the mapping $(Na, Nb) \to Nab^{-1}$ of $G/N \times G/N$ into G/N is continuous. Let V be a neighbourhood of $(a\phi)(b\phi)^{-1}$. Then $V\overset{-1}{\phi}$ is a neighbourhood of ab^{-1} in G, and hence there are neighbourhoods U_1 of a and U_2 of b such that $U_1 U_2^{-1} \subseteq V\overset{-1}{\phi}$. Now $U_1\phi$ and $U_2\phi$ are neighbourhoods of $a\phi$, $b\phi$ respectively and

$$(U_1\phi)(U_2\phi)^{-1} = (U_1 U_2^{-1})\phi \subseteq V,$$

which establishes the continuity. If further, N is closed, then since $(e\phi)\overset{-1}{\phi} = N$, it follows that $\{e\phi\}$ is closed and by Theorem 2.1.3, G/N is a topological group. The final assertion follows from Theorem 2.3.1 and the remark following it.

COROLLARY. *If N is an open normal subgroup of a topological group G, then G/N is discrete.*

For then the cosets of N are open; it follows that each point of G/N is open and hence G/N is discrete.

Suppose now that G and H are any topological groups, and consider a continuous homomorphism θ of G into H. Let N be the kernel of θ: $N = e'\overset{-1}{\theta}$, where e' is the unit element of H. Then N is a normal subgroup of G, and since θ is continuous, N is closed in G. Hence the quotient group G/N with its induced topology becomes a topological group. If ϕ denotes again the natural mapping of G onto G/N, then θ can be written in the form

$$\theta = \phi\bar{\theta}, \tag{4}$$

where $\bar{\theta}$ is an isomorphism of G/N with a subgroup of H. By Theorem 2.3.1, $\bar{\theta}$ is continuous. Thus we have proved

THEOREM 2.3.3. *A continuous homomorphism θ of a topological group G into another topological group H can always be expressed as the result of the natural homomorphism of G onto G/N, where N is the kernel of θ, followed by a continuous isomorphism of G/N into H.*

The mapping $\bar{\theta}$ appearing in (4) is said to be *induced* by the mapping θ. An isomorphism between topological groups which is continuous and whose inverse is also continuous is called a *topological isomorphism.* It is easily verified that the isomorphism $\bar{\theta}$ induced by a homomorphism θ is a topological isomorphism if and only if θ is continuous and open as a mapping of G onto $G\theta$.

In order to show that a homomorphism θ is continuous it is sufficient to show that θ is continuous at e. For if θ is a homomorphism of G into H which is continuous at e, then given any nucleus W of H, we can find a nucleus V of G such that $V\theta \subseteq W$. Hence for any a, $x \in G$, $x \in Va$ implies $x\theta \in W(a\theta)$; this expresses the continuity of θ at a. Thus we have

THEOREM 2.3.4. *A homomorphism of a topological group G is continuous on the whole of G, provided that it is continuous at the unit element.*

Ex. 1. If H is a subgroup of the topological group G, then its topological closure $\bar{H}$ is again a subgroup of G.

Ex. 2. If H is any subgroup of G, then the topology induced on G/H by G is the only topology for which the natural mapping is continuous and open.

Ex. 3. If H is a closed subgroup of G, then G/H is a homogeneous space.

Ex. 4. Let R be the additive group of real numbers with the usual topology defined by $|a-b|$. The subgroup Z consisting of all the integers is closed in R, and the quotient group R/Z is topologically isomorphic to the multiplicative group of complex numbers of absolute value 1, with the same distance topology. Qua topological space, R/Z is a manifold and is homeomorphic with the torus T mentioned in 1.2. Therefore R/Z is often called the *torus group* in one dimension, or also the *additive group of real numbers mod* 1.

Ex. 5. Any homomorphism of a discrete group G onto a topological group H is continuous, but it is not open, unless H is also discrete.

2.4. Connected topological groups. We recall that a topological space T is said to be *connected* if it has no subspace different from $\emptyset$ and T which is both closed and open. If T is any topological space then each point p of T is contained in a unique maximal connected subset K_p of T, which is necessarily closed.† This set K_p is called the *connected component* of p in T, or, more briefly, the p-component of T. In the case of a topological group G, the connected component of e in G is called the e-component, or the *identity component*, of G.

THEOREM 2.4.1. *If G is any topological group, then the identity component of G is a closed normal subgroup K of G, and for any $a \in G$, the a-component of G is the coset Ka.*

Proof. Let K be the e-component of G; then K is closed. If $x \in K$, then K and Kx^{-1} are connected and both contain e, hence $K \cup Kx^{-1}$ is connected and it contains K; therefore, by the maximality of K, we have $K \cup Kx^{-1} = K$, i.e. $Kx^{-1} \subseteq K$. Since this is true for all $x \in K$, we have $KK^{-1} \subseteq K$, i.e. K is a subgroup of G. If $y \in G$, then e belongs to both K and $y^{-1}Ky$, and by a similar argument we conclude that $y^{-1}Ky \subseteq K$, therefore K is normal in G. Finally, if K_a is the a-component of G, then K_a and Ka are connected and contain a, hence $Ka \subseteq K_a$; similarly $K_a a^{-1} \subseteq K$, whence $K_a = Ka$, which is what we wished to prove.

Although, as we have just seen, the identity component of any group is closed, it need not be open. For example, in the additive group of rational numbers with the topology defined by the

† Cf. for example, Bourbaki[1].

metric $|a-b|$ (as in 2.1, illustration 3) the identity component is the subgroup consisting of 0 alone; but this subgroup is not open, since the topology is not discrete.

On the other hand, any open subgroup H of a topological group G is necessarily also closed: If H is open, then so is Ha, for each $a \in G$, and therefore $\bigcup_{a \notin H} Ha$ is open, whence H, its complement in G, must be closed. For open subgroups we have the following version of the second isomorphism theorem:

THEOREM 2.4.2. *Let G be a topological group, H an open subgroup of G and N a normal subgroup of G; then $H/H \cap N$ is topologically isomorphic to HN/N.*†

Proof. Let ϕ be the natural mapping of G onto G/N; by definition of the topology on G/N, this mapping is continuous. When we restrict ϕ to H, the image becomes HN/N and the kernel $H \cap N$. Thus $H/H \cap N$ is algebraically isomorphic to HN/N. Let θ be the isomorphism of $H/H \cap N$ onto HN/N defined in this way, and let ψ be the natural mapping of H onto $H/H \cap N$. Then

$$\phi = \psi\theta.$$

Both ϕ and ψ are continuous and hence, by Theorem 2.3.1, θ is also continuous. It is one-one and so it only remains to show that $\overset{-1}{\theta}$ is also continuous. For this it is enough to prove—by Theorem 2.3.4—that θ maps any nucleus of $H/H \cap N$ onto a nucleus of HN/N. The general nucleus of $H/H \cap N$ has the form $W = (H \cap V)(H \cap N)$, where V is a nucleus of G. Since H is open in G, W is itself a nucleus of G, and so $W\theta = WN/N$ is a nucleus of HN/N. This completes the proof.

Later we shall be dealing with groups in which the identity component is always open, and we can then apply Theorem 2.4.2, taking H to be the identity component of G.

THEOREM 2.4.3. *If G is a connected topological group, then any nucleus V is a system of generators of G, qua abstract group.*

We have to show that the least subgroup of G which contains V is G itself. Denote this subgroup by H, then since $V \subseteq H$, we

† Without the restriction on H the theorem fails to be true. See, for example, Bourbaki[2], p. 15.

have, for any $x \in H$, $Vx \subseteq HH \subseteq H$, i.e. H contains a neighbourhood of each of its points and is therefore open. Hence it is also closed, and since G is connected we must have $H = \emptyset$ or $H = G$. But H contains e, so $H = G$ and the proof is complete.

2.5. Local groups. A *local group* is a set V with the following properties:

L. 1. *V is a Hausdorff space.*

L. 2. *There is a binary operation with values in V*

$$(x, y) \to xy,$$

which is defined for certain pairs of points $x, y \in V$; and a unary operation with values in V

$$x \to x^{-1}$$

defined for certain points $x \in V$.

L. 3. *The mappings $(x, y) \to xy$ and $x \to x^{-1}$ are continuous.*

L. 4. *If $x, y, z \in V$ and $(xy)z$, $x(yz)$ are both defined, then*

$$(xy)z = x(yz).$$

L. 5. *There exists $e \in V$ such that $e^{-1} = e$ and*

$$xe = x \quad \textit{for all } x \in V.$$

L. 6. *If x^{-1} is defined, then so is xx^{-1}, and*

$$xx^{-1} = e.$$

Here L. 3 is taken to mean: If $xy = z$ and W is a neighbourhood of z, then there exist neighbourhoods W' of x and W'' of y such that every product $w'w''$ ($w' \in W'$, $w'' \in W''$) is defined and belongs to W. Similarly, if x^{-1} is defined and W is a neighbourhood of x^{-1}, then there is a neighbourhood U of x such that u^{-1} is defined and belongs to W for all $u \in U$. The element e, which is uniquely defined by L. 5–6, is called the *unit element* of V, and any neighbourhood of e is called a *nucleus* of V.

Two local groups V and V' are said to be *topologically isomorphic*, if there is a homeomorphism $x \leftrightarrow x'$ between V and V' such that the product xy in V is defined if and only if the product $x'y'$ of the corresponding elements in V' is defined, and in this case

$$(xy)' = x'y'.$$

It is clear that in a topological group, any open nucleus is a local group, and the definitions of 'unit element' and 'nucleus' just given for local groups then agree with the definitions of these terms for groups. However, it is not true that a local group can always be embedded in a topological group, nor need it have a nucleus which is isomorphic to a nucleus of a topological group.† Nevertheless, local groups share a number of properties with open nuclei of topological groups which we list in Theorem 2.5.1 below. We could, of course, dispense with this theorem by taking it as a basis for the definition of local groups (and for this reason the reader may well omit it), but it makes it a little simpler to verify that a given structure is a local group.

THEOREM 2.5.1. *Let V be a local group, then there is a nucleus U such that*

(i) $(xy)z = x(yz)$ *for all* $x, y, z \in U$.

(ii) $x^{-1}x = xx^{-1} = e$ *for all* $x \in U$.

(iii) $ex = xe = x$ *for all* $x \in U$.

(iv) $U^{-1} = U$ *and* $(x^{-1})^{-1} = x$ *for all* $x \in U$.

Conditions (i)–(iv) are understood as implying that all the products and inverses written down are actually defined. Any nucleus of V satisfying (i)–(iv) is called a *germ* of the local group V.

Proof. By L. 5, $ee = e$, and hence (by L. 3) there exists a nucleus U_1 such that $U_1U_1 \subseteq V$ and a nucleus U_2 such that $U_2U_2 \subseteq U_1$. Further, since $e^{-1} = e$, there is a nucleus U_3 such that $U_3^{-1} \subseteq U_2$, and a nucleus U_4 such that $U_4^{-1} \subseteq U_3$. Put $U_5 = U_2 \cap U_3 \cap U_4$, then U_5 is a nucleus and

$$U_5U_5 \subseteq U_1, \quad U_5 \subseteq U_2, \quad U_5^{-1} \subseteq U_2, \quad (U_5^{-1})^{-1} \subseteq U_1, \tag{5}$$

since U_5 is contained in U_2, U_2, U_3, U_4 respectively. Further, we have $U_5^{-1}U_5 \subseteq U_2U_2 \subseteq U_1$, so that

$$U_5^{-1}U_5 \subseteq U_1. \tag{6}$$

Next we choose a nucleus U_6 such that $U_6^{-1} \subseteq U_5$, a nucleus U_7 such that $U_7^{-1} \subseteq U_6 \cap U_5$ and put $U = U_7 \cap U_7^{-1}$. Then U is a nucleus with the required properties. We first show that U_5 satisfies (i)–(iii).

† Cf. A. I. Malcev[10].

(i) follows since $(U_5U_5)(U_5U_5) \subseteq U_1U_1 \subseteq V$. To prove (ii), let $x \in U_5$, then x^{-1} is defined and hence $xx^{-1} = e$ by L. 6. Moreover, $x^{-1}x \in U_5^{-1}U_5 \subseteq U_1$ by (6), hence

$$(x^{-1}x)\,x^{-1} = x^{-1}(xx^{-1}) = x^{-1}e = x^{-1}$$

(using L. 5). Write $y = x^{-1}x$, then $y \in U_1$ and $yx^{-1} = x^{-1}$; we have to show that $y = e$. By (5), $(x^{-1})^{-1} \in U_1$, and

$$e = x^{-1}(x^{-1})^{-1} = (yx^{-1})\,(x^{-1})^{-1} = y(x^{-1}(x^{-1})^{-1}) = ye = y,$$

i.e. $x^{-1}x = e$. Turning to (iii), we have $xe = x$ for all $x \in V$, by L. 5; moreover, if $x \in U_5$, then by (ii), $ex = (xx^{-1})\,x = x(x^{-1}x) = xe = x$. Thus U_5 satisfies (i)–(iii). Clearly any subset of U_5, in particular U, also satisfies (i)–(iii).

To show that U also satisfies (iv), let $x \in U_7$. Then $x^{-1} \in U_6 \cap U_5$; hence by (ii), applied to U_5, $(x^{-1})^{-1} \in U_1$ and $x^{-1}(x^{-1})^{-1} = e$. Therefore

$$x = xe = x[x^{-1}(x^{-1})^{-1}] = (xx^{-1})\,(x^{-1})^{-1} = e(x^{-1})^{-1} = (x^{-1})^{-1},$$

because $(x^{-1})^{-1} \in U_5$. This proves that

$$(x^{-1})^{-1} = x \tag{7}$$

for all $x \in U_7$, and since $U \subseteq U_7$, (7) holds for all $x \in U$. It also follows from (7) that $(U_7^{-1})^{-1} = U_7$. Hence

$$U^{-1} = (U_7 \cap U_7^{-1})^{-1} = U_7^{-1} \cap U_7 = U.$$

This completes the proof.

Two topological groups are said to be *locally isomorphic* if they have open nuclei which, qua local groups, are topologically isomorphic.

Any two topological groups which are topologically isomorphic are clearly locally isomorphic, but the converse is not true. Consider, for example, R, the additive group of real numbers, and R/Z, the quotient of R by the group of integers (cf. 2.3, Ex. 4). The natural mapping of R onto R/Z is continuous and open; if we restrict it to the interval $-\frac{1}{3} < x < \frac{1}{3}$, then it is a homeomorphism, and it is easily verified that it provides a local isomorphism between R and R/Z. However, the mapping is not an isomorphism; in fact R and R/Z are not even isomorphic as abstract groups, since R/Z, but not R, contains elements of finite order besides the unit element.

Ex. 1. The definition of topological isomorphism given for local groups can also be used for (not necessarily open) nuclei of topological groups. Show that if two topological groups have topologically isomorphic nuclei, then they have also topologically isomorphic *open* nuclei, and hence are locally isomorphic.

Ex. 2. Show that an isomorphism of one topological group G onto another, H, which is continuous and whose restriction to some nucleus of G is a homeomorphism, need not be a local isomorphism. (Consider the mapping $n \to \alpha n \pmod 1$ for irrational α.)

2.6. Lie groups

DEFINITION. A *Lie group* is a set $\mathfrak{G}$ such that

LG. 1. *$\mathfrak{G}$ is a group.*

LG. 2. *$\mathfrak{G}$ is an analytic manifold.*

LG. 3. *The mapping $(x, y) \to xy$ of the product manifold $\mathfrak{G} \times \mathfrak{G}$ into $\mathfrak{G}$ is analytic.*

Let $\mathfrak{G}$ be any Lie group and choose a fixed chart at e. If V is the nucleus on which it is defined, we denote the coordinates of $x \in V$ by x^i $(i = 1, \ldots, n)$. By LG. 3, there is a nucleus W such that

$$(xy)^i = \phi^i(x^1, \ldots, x^n, y^1, \ldots, y^n) \quad (x, y \in W), \tag{8}$$

where the ϕ^i are analytic functions of their arguments. In fact, it is enough to choose a nucleus W such that $WW \subseteq V$. The functions ϕ^i are called the *composition functions* of $\mathfrak{G}$ (in the given chart); we shall usually write (8) as $(xy)^i = \phi^i(x, y)$ in the shorter notation of 1.1.

Conversely, if in a topological group there is a chart at e in which the product, when expressed in terms of its arguments, as in (8), is analytic, then the group can be defined as a Lie group. But before proving this we must show that a Lie group can indeed always be regarded as a topological group.

THEOREM 2.6.1. *In a Lie group $\mathfrak{G}$ the mapping $x \to x^{-1}$ is analytic.*

Proof. We take a chart at e and denote the coordinates of $x \in G$ by x^i. Then in some nucleus, $(xy)^i = \phi^i(x, y)$, where the ϕ^i are analytic and $\phi^i(e, y) = y^i$. Hence

$$\left(\frac{\partial \phi^i}{\partial y^j}\right)_{x=y=e} = \delta^i_j,$$

and by Theorem A 4 of the appendix, the equations $(xy)^i = e^i$ have a unique solution $y^i = \theta^i(x)$ for all x in some nucleus, where the θ^i are analytic at e. But $y = x^{-1}$ when $xy = e$, and so the mapping $\tau : x \to x^{-1}$ is analytic at e. To prove that it is analytic at any point a of $\mathfrak{G}$, let $x = au$, then by applying in turn $\lambda_{a^{-1}}, \tau, \rho_{a^{-1}}$, we obtain $x = au \to u \to u^{-1} \to u^{-1}a^{-1} = x^{-1}$; the translations are analytic mappings by LG. 3, and therefore τ is analytic at a. This completes the proof.

Since a manifold is always a Hausdorff space and an analytic mapping is necessarily continuous, we have the

COROLLARY. *Any Lie group is a topological group with respect to the topology induced by its analytic structure.*

THEOREM 2.6.2. *Let G be a topological group and suppose that there is a chart (x): $x \to (x^i)$ defined on a nucleus V. If the product xy, when expressed in terms of this chart, is analytic,*

$$(xy)^i = \phi^i(x, y) \quad (\phi^i \text{ analytic}), \tag{9}$$

then G can be defined in just one way as a Lie group $\mathfrak{G}$ such that the given chart—when restricted to a suitable nucleus—belongs to the analytic structure of $\mathfrak{G}$, and the topology of $\mathfrak{G}$ then coincides with the given topology on G.

Proof. We first define the analytic structure on G. In order to do this we choose a nucleus W such that $WW \subseteq V$. Next we choose a nucleus W_1 such that $W_1^{-1}W_1 \subseteq W$; then $W_1 \subseteq W \subseteq V$ and for any $a \in G$, $W_1 a$ is a neighbourhood of a. We define a chart on $W_1 a$ by the rule: If $u \in W_1 a$, say $u = xa$, $x \in W_1$, then

$$(u)^i_a = x^i, \tag{10}$$

where x^i are the coordinates of x in the chart given on V. Since the correspondence $x \to xa$ is a homeomorphism, it follows that the correspondence $u \to (x^i)$ is a homeomorphism, so that (10) defines a chart on $W_1 a$. In this way a chart is defined at each point of G. To show that all these charts are analytically related, suppose that $W_1 a \cap W_1 b \neq \emptyset$, say $u \in W_1 a \cap W_1 b$. Let $u = xa = yb$, where $x, y \in W_1$. Then

$$ba^{-1} = y^{-1}x \in W_1^{-1}W_1 \subseteq W, \quad \text{and} \quad ab^{-1} = x^{-1}y \in W_1^{-1}W_1 \subseteq W,$$

hence
$$(u)_a^i = x^i = (yba^{-1})^i = \phi^i(y, ba^{-1}),$$
$$(u)_b^i = y^i = (xab^{-1})^i = \phi^i(x, ab^{-1}).$$

These equations, which give the transformation of coordinates from x^i to y^i and in the reverse direction, are analytic by (9). Therefore we have an analytic family of charts covering G, and by Theorem 1.2.1, this defines an analytic structure on G; by definition the original chart, restricted to W_1, belongs to this structure. Moreover, by (10) there is for any point a of the resulting analytic manifold $\mathfrak{G}$, an analytic homeomorphism mapping a neighbourhood of a onto a neighbourhood of e; hence it is enough to verify local properties at e. The operation of multiplication is analytic at (e, e), by (9), and hence everywhere; thus $\mathfrak{G}$ is a Lie group. The mapping of G onto $\mathfrak{G}$ which maps each point onto itself is a homeomorphism at e, and hence everywhere; in other words, the topology of $\mathfrak{G}$ coincides with the given topology on G. Finally, the analytic structure is unique, since it is determined near e by the given chart. This completes the proof.

COROLLARY. *The dimension of a Lie group is the same at all its points.*

For the right translations are analytic homeomorphisms and any point may be mapped into any other point in this way. The corollary therefore follows from Theorem 1.8.2.

By the corollary we may speak of the *dimension* of a Lie group.

Theorem 2.6.2 can be generalized considerably. Thus it is usual to suppose that the functions $\phi^i(x, y)$ are only twice continuously differentiable; with this hypothesis it is still possible to develop the theory and eventually show that the group defined in this way possesses an analytic structure with respect to which it is a Lie group (cf. Mayer and Thomas [11]). If no special assumption is made about the ϕ's, then all that is given about them is that they are continuous functions, because G is a topological group. Under these assumptions it is still true that an analytic structure can be defined on G which turns it into a Lie group. This is a deep result due to Montgomery, Gleason and Zippin (cf. Montgomery and Zippin [12]); it constitutes the solution of a problem raised by Hilbert in 1900 and known as Hilbert's fifth problem. We shall not prove the existence of Lie group structures under these weaker hypotheses, but confine ourselves to the analytic case.

A further point arising from Theorem 2.6.2 is the following: The theorem shows that a chart at e (with analytic composition functions) in a topological group G determines a unique Lie group structure on G, but it does not say whether different Lie group structures can be determined on G in this way, starting only with a given topology. If in a topological group G two charts at e are given, such that in each of them the composition functions are analytic, then they will each, by Theorem 2.6.2, define an analytic structure on G, and we may ask whether these structures are necessarily the same. Clearly the structures will coincide if and only if the two charts are analytically related; we shall see later (Theorem 6.3.3) that this is always the case so that in a Lie group the analytic structure is already determined by the topology.

Examples of Lie groups

1. $\mathfrak{R}$ or more generally $\mathfrak{R}^n$. The group structure of $\mathfrak{R}^n$ is that of an n-dimensional vector space over the field R. The analytic structure is as defined in 1.2. A chart is defined by the coordinates of a vector v with respect to a fixed basis: $v = x^i v_i$. Then $\mathfrak{R}^n$ is a Lie group since the multiplication law can be written

$$(x+y)^i = x^i + y^i.$$

2. $\mathfrak{T}$ or more generally $\mathfrak{T}^n$. Near the origin (i.e. the unit element of $\mathfrak{T}^n$) the multiplication law is the same as for $\mathfrak{R}^n$, hence $\mathfrak{T}^n$ is also a Lie group. It is locally isomorphic to $\mathfrak{R}^n$, but not isomorphic to it (see 2.5).

3. The set $\mathfrak{R}_n$ of all $n \times n$ matrices with coefficients in R: the general element is $a = (a_{ij})$. With respect to addition $\mathfrak{R}_n$ is a Lie

group of dimension n^2 (see example 1). The set of non-singular matrices is an open subset, for its complement—being the inverse image of 0 under the continuous mapping $a \to \det a$—is closed. So this set can be defined as a submanifold of $\mathfrak{R}_n$, by Theorem 1.9.2. With respect to matrix multiplication this represents the general linear group $GL(n, R)$ (see 1.2) and the multiplication is given by

$$(ab)_{ij} = a_{ik} b_{kj}. \dagger$$

4. The rotations about the origin in 3-dimensional space form a Lie group. It may be represented by the proper orthogonal matrices, and from this point of view it forms a subgroup of the $GL(3, R)$. More generally, the orthogonal $n \times n$ matrices form a Lie group.

5. Every discrete group is a Lie group. For then e has the neighbourhood consisting of e alone, which may be regarded as being homeomorphic to $R^0 = \{0\}$. Thus every discrete group is a zero-dimensional Lie group. Conversely, a zero-dimensional Lie group has a nucleus homeomorphic to R^0 and hence is discrete.

Ex. The composition function $\phi(x, y) = (x^3 + y^3)^{\frac{1}{3}}$ defines the set R of real numbers (with the usual topology) as a topological group, but not as a Lie group, because ϕ is not analytic.

2.7. Local Lie groups. In analogy with local topological groups we define a local Lie group by abstraction from an open nucleus in a Lie group. For convenience we assume the nucleus to be connected. Thus we have the following

DEFINITION. A *local Lie group* is a set $\mathfrak{B}$ such that

LL. 1. $\mathfrak{B}$ is a connected analytic manifold.

LL. 2. $\mathfrak{B}$ is a local group with respect to the topology defined by LL. 1.

LL. 3. The mapping $(x, y) \to xy$ defined on $\mathfrak{U} \times \mathfrak{U}$, where $\mathfrak{U}$ is an open germ of $\mathfrak{B}$, is analytic.‡

We note the special case of this definition where there exists a single chart covering the whole of $\mathfrak{B}$. If we identify $\mathfrak{B}$ with the

† The $GL(n, R)$ is a typical example of an *algebraic group* and it forms the starting point for a theory of Lie groups over any field of characteristic zero. Cf C. Chevalley[6].

‡ The set $\mathfrak{U}$, being open, is a submanifold of $\mathfrak{B}$, by Theorem 1.9.2.

region of R^n to which it corresponds, then the definition of a local Lie group in this case may be stated as follows:

There is a set of n real functions $\phi^i(x, y)$ of $2n$ real variables x^i, y^i $(i = 1, \ldots, n)$, analytic at $x^i = 0, y^i = 0$, and such that

$$\phi^i(x, \phi(y, z)) = \phi^i(\phi(x, y), z), \tag{11}$$

$$\phi^i(x, 0) = x^i, \tag{12}$$

$$\det\left(\frac{\partial \phi^i}{\partial y^j}\right)_{x=y=0} \neq 0, \tag{13}$$

for x, y, z in some neighbourhood of $O = (0, 0, \ldots, 0)$.

For if such functions are defined, then by (13), the equations $\phi^i(x, y) = 0$ can be solved uniquely for y in terms of x in some neighbourhood of O:

$$y^i = \theta^i(x). \tag{14}$$

The functions $\phi^i(x, y)$ and $\theta^i(x)$ are defined on some neighbourhood W of O and are analytic there. Let us denote the points of R^n with coordinates $\phi^i(x, y)$ and $\theta^i(x)$ by xy and x^{-1} respectively, and write e instead of O. Then the mappings $(x, y) \to xy$ and $x \to x^{-1}$ are continuous (in fact analytic) mappings of $W \times W$ and W respectively into R^n. They satisfy the equations

$$x(yz) = (xy)\,z, \quad xe = x, \quad xx^{-1} = e,$$

and since e has a unique inverse, $e^{-1} = e$. If we restrict the functions ϕ^i and θ^i so that their values lie in the e-component of the interior of W, we obtain a local Lie group with the analytic composition functions ϕ^i.

Two Lie groups are said to be *analytically isomorphic* if there is an isomorphism between them which is also an analytic homeomorphism; they are *locally analytically isomorphic* if there is a local isomorphism between them which together with its inverse is analytic.

In 2.6 we saw that a Lie group is determined by its structure as a topological group, provided that its analytic structure is given near e. We shall now see that even when no topology is given, beyond that of the analytic structure near e, we can topologize the group in just one way so that it becomes a Lie group.

THEOREM 2.7.1. *Let G be an abstract group and $\mathfrak{B}$ a subset of G such that*

(i) *$\mathfrak{B}$ generates G,*

(ii) *$\mathfrak{B}$ is a local Lie group with respect to the multiplication of G.*

Then G can be defined in one and only one way as a Lie group $\mathfrak{G}$ such that the group structure of $\mathfrak{G}$ coincides with the given group structure of G and the analytic structure coincides with the given analytic structure on some nucleus of $\mathfrak{B}$. Moreover, $\mathfrak{G}$ is connected.

Proof. Suppose that we can define a topology on G which turns it into a topological group and induces on $\mathfrak{B}$ the topology given by the analytic structure. Then the analytic structure of $\mathfrak{B}$ can be extended to the whole of G in such a way that G becomes a Lie group, by Theorem 2.6.2.

We define such a topology on G as follows: Consider the family $\mathscr{V}$ of nuclei of $\mathfrak{B}$ and let $\mathscr{W}$ be the family of subsets W of G such that $W \cap \mathfrak{B} \in \mathscr{V}$. In particular, every nucleus itself belongs to $\mathscr{W}$, i.e. $\mathscr{V} \subseteq \mathscr{W}$. We show first that $\mathscr{W}$ satisfies the five axioms of Theorem 2.2.1. V.1–3 are immediate from the corresponding properties of $\mathscr{V}$. To prove V.4, let $W \in \mathscr{W}$, then $W \cap \mathfrak{B} \in \mathscr{V}$, and since $\mathfrak{B}$ is a local group, there is a $V_1 \in \mathscr{V}$ such that

$$V_1 V_1^{-1} \subseteq W \cap \mathfrak{B};$$

hence $V_1 V_1^{-1} \subseteq W$ and $V_1 \in \mathscr{W}$. Next, to prove V.5, we suppose again that $W \in \mathscr{W}$ and $a \in G$. Since $\mathfrak{B}$ generates G, we have $a = a_1 a_2 \ldots a_k$, where $a_i \in \mathfrak{B}$, or $\mathfrak{B}^{-1}$. Now $a_i e a_i^{-1} = e$, therefore by the continuity of the multiplication, there is a $U_1 \in \mathscr{V}$ such that $a_1 U_1 a_1^{-1} \subseteq W$. Similarly

there exists $U_2 \in \mathscr{V}$ such that $a_2 U_2 a_2^{-1} \subseteq U_1$,

....

there exists $U_k \in \mathscr{V}$ such that $a_k U_k a_k^{-1} \subseteq U_{k-1}$.

Hence

$$\begin{aligned} a_1 \ldots a_k U_k a_k^{-1} \ldots a_1^{-1} &\subseteq a_1 \ldots a_{k-1} U_{k-1} a_{k-1}^{-1} \ldots a_1^{-1} \\ &\ldots\ldots\ldots\ldots\ldots\ldots \\ &\subseteq a_1 U_1 a_1^{-1} \\ &\subseteq W. \end{aligned}$$

Thus $aU_k a^{-1} \subseteq W$, i.e. $U_k \subseteq a^{-1} W a$, and this shows that

$$a^{-1} W a \in \mathscr{W}.$$

By Theorem 2.2.1, G can therefore be defined as a topological group with $\mathscr{W}$ as system of nuclei. Now in $\mathfrak{V}$, every nucleus belongs to $\mathscr{W}$, and conversely, every set of $\mathscr{W}$ contains a set of $\mathscr{V}$ which is therefore a nucleus of $\mathfrak{V}$. Therefore the topology defined on G by $\mathscr{W}$ coincides on $\mathfrak{V}$ with the manifold topology. By Theorem 2.6.2, G can now be defined as a Lie group $\mathfrak{G}$ and its analytic structure is unique, since it is given near e.

The group $\mathfrak{G}$ is connected: Let $W = \mathfrak{V} \cup \mathfrak{V}^{-1}$, then W, as the union of the two connected sets $\mathfrak{V}$, $\mathfrak{V}^{-1}$ with the common point e, is again connected. Each set W^n† as a continuous image of $W \times W \times \ldots \times W$, is connected, and the sets $W, W^2, \ldots$ have the common point e, therefore their union $\bigcup_{n=1}^{\infty} W^n$ is connected; but this union is equal to $\mathfrak{G}$, since $\mathfrak{V}$ generates $\mathfrak{G}$; thus the proof is complete.

The following example shows that condition (i) cannot be omitted from the theorem: Let G be the group of rotations about the origin in R^3, and V the subgroup of rotations about the z-axis. The group V is isomorphic to the torus and may therefore be defined as a 1-dimensional Lie group. If we now take V as a coordinate neighbourhood of e, as in the proof of Theorem 2.7.1, then V is open; any subgroup conjugate to V is likewise open, thus the subgroup of rotations about the x-axis is open and this subgroup meets V in $\{e\}$. Therefore $\{e\}$ is an open subgroup and the topology defined on G in this way is the discrete topology. But this induces the discrete topology on V, whereas the initial topology on V was not discrete.

2.8. Analytic subgroups of a Lie group

Definition. If $\mathfrak{G}$ is a Lie group, an *analytic subgroup* of $\mathfrak{G}$ is a subset $\mathfrak{H}$ of $\mathfrak{G}$ such that

(i) *$\mathfrak{H}$ is a subgroup of $\mathfrak{G}$ (qua abstract group),*

(ii) *$\mathfrak{H}$ is a submanifold of $\mathfrak{G}$ (qua analytic manifold).*

An analytic subgroup of a Lie group need not be closed or even a subspace, but it is always a Lie group, as we shall now prove.

† W^n stands for $WW \ldots W$ (n factors).

THEOREM 2.8.1. *Any analytic subgroup of a Lie group $\mathfrak{G}$ is a Lie group. Conversely, a Lie group $\mathfrak{H}$ which is a subgroup of $\mathfrak{G}$ is an analytic subgroup of $\mathfrak{G}$, if at the unit element e of $\mathfrak{G}$ there exist charts (x) in $\mathfrak{G}$ and (u) in $\mathfrak{H}$, such that*

$$x^i \mid \mathfrak{H} = \chi^i(u) \quad (\chi^i \text{ analytic, rank } (\partial\chi^i/\partial u^\rho)_e = \dim \mathfrak{H}). \quad (15)$$

Proof. Let $\mathfrak{H}$ be an analytic subgroup of $\mathfrak{G}$ and let $a, b \in \mathfrak{H}$. Then $ab \in \mathfrak{H}$ and we may choose coordinates $z^1, \ldots, z^n$ at ab in $\mathfrak{G}$ such that $\bar{z}^1, \ldots, \bar{z}^\nu$ (where $\bar{z}^i = z^i \mid \mathfrak{H}$ and $\nu = \dim \mathfrak{H}$) form a chart at ab in $\mathfrak{H}$. Denote the coordinates of any point z near ab by z^i; then xy is near ab when x and y are near a and b respectively, and $(xy)^i$ is analytic in x and y. This remains true when x and y are restricted to $\mathfrak{H}$ and hence, for any $f \in \mathscr{A}_{ab}(\mathfrak{H})$, $f(\overline{(xy)}^1, \ldots, \overline{(xy)}^\nu)$ is analytic. Thus the mapping $(x, y) \to xy$, restricted to $\mathfrak{H} \times \mathfrak{H}$, is analytic and so $\mathfrak{H}$ is a Lie group.

Conversely, if $\mathfrak{H}$ is a Lie group contained as a subgroup in $\mathfrak{G}$, with charts at e satisfying (15), then we can translate the charts to any point of $\mathfrak{H}$ by a right translation, which is necessarily an analytic homeomorphism, whether it is considered in $\mathfrak{G}$ or restricted to $\mathfrak{H}$. Hence (15) holds at all points of $\mathfrak{H}$ and so, by Theorem 1.9.1, $\mathfrak{H}$ is a submanifold of $\mathfrak{G}$. Therefore $\mathfrak{H}$ is an analytic subgroup of $\mathfrak{G}$, as we wished to prove.

It is sometimes convenient to have a version of this theorem in terms of local groups.

THEOREM 2.8.2. *Let $\mathfrak{G}$ be a Lie group and $\mathfrak{B}$ a subset of $\mathfrak{G}$ which is a local Lie group. If the multiplication in $\mathfrak{B}$ agrees with the multiplication in $\mathfrak{G}$, and if at e there are charts (x) in $\mathfrak{G}$ and (u) in $\mathfrak{B}$, such that*

$$x^i \mid \mathfrak{B} = \chi^i(u),$$

where the χ^i are analytic at e and the rank of $(\partial\chi^i/\partial u^\rho)_e$ equals the dimension of $\mathfrak{B}$ at e, then the subgroup of $\mathfrak{G}$ generated by $\mathfrak{B}$ is a connected analytic subgroup of $\mathfrak{G}$.

For the proof we need only observe that the subgroup generated by $\mathfrak{B}$ can be defined as a connected Lie group in such a way that it induces the given analytic structure on a suitable nucleus of $\mathfrak{B}$ (Theorem 2.7.1) and we may therefore apply Theorem 2.8.1 to complete the proof.

We note that if $\mathfrak{G}$ is a connected Lie group of dimension n, then the dimension of any proper analytic subgroup $\mathfrak{H}$ of $\mathfrak{G}$ is less than n. For otherwise there would be a nucleus V of $\mathfrak{G}$ which is contained in $\mathfrak{H}$. Since V generates $\mathfrak{G}$, by Theorem 2.4.3, it would follow that $\mathfrak{H} = \mathfrak{G}$, which contradicts the hypothesis that $\mathfrak{H}$ is proper.

This property need not hold when $\mathfrak{G}$ is not connected. For example, let Z be the additive group of integers, then $R \times Z$ can be regarded as a 1-dimensional analytic subgroup of R^2, and we have the inclusion $R \subset R \times Z \subset R^2$; here the dimensions are 1, 1, 2 respectively. From the point of view of the local theory the difference between R and $R \times Z$ is rather inessential; this is expressed more precisely in

THEOREM 2.8.3. *If $\mathfrak{G}$ is any Lie group and $\mathfrak{H}$ a connected subgroup containing a nucleus of $\mathfrak{G}$, then $\mathfrak{H}$ coincides with $\mathfrak{G}_0$, the identity component of $\mathfrak{G}$. This component $\mathfrak{G}_0$ is open in $\mathfrak{G}$ and the quotient $\mathfrak{G}/\mathfrak{G}_0$ is discrete; moreover, $\mathfrak{G}$ induces an analytic structure on $\mathfrak{G}_0$ for which $\mathfrak{G}_0$ is an analytic subgroup of $\mathfrak{G}$ of the same dimension as $\mathfrak{G}$.*

These facts all follow readily from previous results: Let $\mathfrak{G}_0$ be the e-component of $\mathfrak{G}$, then $\mathfrak{G}_0 \supseteq \mathfrak{H}$, because $\mathfrak{H}$ is connected, but by hypothesis $\mathfrak{H}$ contains a nucleus of $\mathfrak{G}$ which, by Theorem 2.4.3, generates $\mathfrak{G}_0$; therefore $\mathfrak{H} \supseteq \mathfrak{G}_0$, which proves that $\mathfrak{H} = \mathfrak{G}_0$. Now the Lie group $\mathfrak{G}$ has a connected nucleus V, and as in the proof of Theorem 2.7.1, we can show that V generates a connected subgroup of $\mathfrak{G}$; by what has been proved, this subgroup must be $\mathfrak{G}_0$, the e-component of $\mathfrak{G}$. Therefore $\mathfrak{G}_0$ contains a neighbourhood V of e. It follows that $\mathfrak{G}_0$ contains a neighbourhood of each of its points: if $x \in \mathfrak{G}_0$, then $Vx \subseteq \mathfrak{G}_0\mathfrak{G}_0 \subseteq \mathfrak{G}_0$. Hence $\mathfrak{G}_0$ is open and by Theorem 2.3.2, Corollary, $\mathfrak{G}/\mathfrak{G}_0$ is discrete. By Theorem 1.9.2, $\mathfrak{G}$ may be defined as a submanifold, and hence as an analytic subgroup of $\mathfrak{G}$, of the same dimension as $\mathfrak{G}$.

This Theorem 2.8.3 shows that in a Lie group the topological interest resides in the identity component; for this reason we can often restrict our attention to connected Lie groups, especially when dealing with local properties. It also makes it possible to use the topological form of the second isomorphism theorem (Theorem 2.4.2) to compare the quotients of $\mathfrak{G}$ with the quotients of $\mathfrak{G}_0$.

Ex. 1. Every discrete subgroup of a Lie group is analytic.

Ex. 2. Every discrete subgroup of a Lie group is closed. (We shall see later that every closed subgroup of a Lie group is analytic.)

Ex. 3. Consider the torus group $\mathfrak{T}^2$ defined in 2.6, with coordinates (x, y) say, where x and y are real numbers mod 1 (cf. 1.9, illustration 2). Then the equations $x = t$, $y = \alpha t$ (mod 1), where t is a real parameter and α is a fixed real number, define a 1-dimensional submanifold of $\mathfrak{T}^2$ which is a subgroup, and therefore an analytic subgroup of $\mathfrak{T}^2$. But if α is irrational, this subgroup is not closed and its topology is not that induced by $\mathfrak{T}^2$.

2.9. One-dimensional Lie groups. The simplest non-trivial example of a Lie group is a 1-dimensional Lie group; we shall consider this case here as an illustration and for later applications.

Let $\mathfrak{G}$ then be a 1-dimensional Lie group; on some nucleus V there is a chart, which assigns to each $x \in V$ a real number x^1, and the composition function ϕ given by the equation

$$(xy)^1 = \phi(x^1, y^1)$$

is analytic. Let us suppose the chart so chosen that $e^1 = 0$, and let us also drop the superscript ^{1}, so that $x, y, \ldots$ are real numbers representing points in $\mathfrak{G}$.

The associative law in $\mathfrak{G}$ is expressed by the equation

$$\phi(\phi(x, y), z) = \phi(x, \phi(y, z)).$$

If we differentiate both sides with respect to z and put $z = 0$, we obtain

$$\left[\frac{\partial \phi(\phi(x, y), z)}{\partial z}\right]_{z=0} = \frac{\partial \phi(x, y)}{\partial y}\left[\frac{\partial \phi(y, z)}{\partial z}\right]_{z=0}.$$

This equation may be written

$$\frac{\partial \phi(x, y)}{\partial y} = \frac{\psi(\phi(x, y))}{\psi(y)}, \tag{16}$$

where

$$\psi(u) = \left[\frac{\partial \phi(u, v)}{\partial v}\right]_{v=0}. \tag{17}$$

We note that since $\phi(0, v) = v$, we have $\psi(0) = 1$. Equation (16) may be written in a rather more suggestive form: Let us put $z = \phi(x, y)$ and denote by δz the change in z due to a small change δy in y, while x is kept fixed. Then (16) states that

$$\frac{\delta z}{\psi(z)} = \frac{\delta y}{\psi(y)}.$$

This equation suggests that we introduce $\int \frac{dy}{\psi(y)}$ as a new coordinate in our group. So we define a mapping $x \to \bar{x}$ by the rule

$$\bar{x} = f(x) \equiv \int_0^x \frac{dt}{\psi(t)}. \tag{18}$$

Clearly $f(x)$ is analytic at $x = 0$ and it has an analytic inverse at $x = 0$, since $\left[\frac{df}{dx}\right]_{x=0} = \frac{1}{\psi(0)} = 1$, so that (18) really defines an admissible chart $(\bar{x})$ at e. Let the composition function in this chart be χ. Thus

$$\bar{z} = \chi(\bar{x}, \bar{y}), \quad \text{where} \quad z = \phi(x, y).$$

In terms of the old chart this equation reads

$$f(\phi(x, y)) = \chi(f(x), f(y)).$$

If we differentiate this equation with respect to y, we find

$$f'(z) \frac{\partial \phi(x, y)}{\partial y} = \frac{\partial \chi}{\partial \bar{y}} f'(y).$$

Now put $y = 0$; then $\bar{y} = 0$, $z = x$ and hence

$$f'(x)\,\psi(x) = \left[\frac{\partial \chi}{\partial \bar{y}}\right]_{\bar{y}=0} f'(0).$$

But $f'(x) = 1/\psi(x)$ and $f'(0) = 1$, by (18); hence $[\partial\chi/\partial\bar{y}]_{\bar{y}=0} = 1$, and therefore, by (16), applied to the new chart,

$$\frac{\partial \chi(\bar{x}, \bar{y})}{\partial \bar{y}} = 1.$$

If we integrate this equation from 0 to $\bar{y}$, remembering that $\chi(\bar{x}, 0) = \bar{x}$, we find that $\chi(\bar{x}, \bar{y}) = \bar{x} + \bar{y}$. This proves the first part of

THEOREM 2.9.1. *In any 1-dimensional Lie group there is a chart* $(\bar{x})$ *at e in which the multiplication is defined by*

$$\overline{xy} = \bar{x} + \bar{y}. \tag{19}$$

If (x^*) *is any chart satisfying the equation*

$$(xy)^* = x^* + y^*, \tag{20}$$

then $$x^* = \lambda\bar{x}, \quad \textit{where} \quad \lambda \neq 0; \tag{21}$$

conversely, any function x^* *defined by* (21) *is a chart at e satisfying* (20).

Any chart with the property expressed in (19) will be called a *canonical chart* and its coordinate a *canonical coordinate*. We shall see later how a certain type of 'canonical' chart can be defined in any Lie group, with a property which reduces to (19) in the case of a 1-dimensional Lie group.

To prove the second part of the theorem, let a chart x^* satisfying (20) be given by

$$x^* = f(\bar{x}).$$

Then $(xy)^* = f(\overline{(xy)}) = f(\bar{x}) + f(\bar{y})$, and hence

$$f(\bar{x} + \bar{y}) = f(\bar{x}) + f(\bar{y}), \tag{22}$$

for all real $\bar{x}$, $\bar{y}$ near 0. Putting $\bar{x} = \bar{y} = 0$ in (22), we find that $f(0) = 0$. Now $f(\bar{x})$ is an analytic function; differentiating (22) with respect to $\bar{y}$ and putting $\bar{y} = 0$, we obtain

$$f'(\bar{x}) = f'(0) = \lambda \quad \text{say},$$

and on integrating this equation we find

$$f(\bar{x}) = \lambda\bar{x},$$

because $f(0) = 0$. Thus $x^* = \lambda\bar{x}$, where $\lambda \neq 0$, since it must be possible to solve this equation for $\bar{x}$ in terms of x^*. Conversely, any x^* given by (21) clearly defines a chart at e which satisfies (20) and so the theorem is established.

The first part of Theorem 2.9.1 may also be expressed by saying that every 1-dimensional Lie group is locally isomorphic to R. In fact (19) expresses just that the mapping $x \to \bar{x}$ of the appropriate nucleus into R is a homomorphism. Since we may take $\bar{x}$ itself to be the coordinate of x, this homomorphism is analytic and is invertible at e. Thus we have a local isomorphism, and this can be extended to a homomorphism defined on the whole of R, by means of

LEMMA 2.9.2. *Let G be any group and $t \to x(t)$ a mapping of some interval $[-\delta, \delta]$ $(\delta > 0)$ of the real line into G such that*

$$x(t)\,x(t') = x(t+t') \tag{23}$$

whenever t, t' and $t+t'$ lie in $[-\delta, \delta]$. Then the mapping can be extended to a homomorphism of the additive group R into G.

Proof. From (23) we deduce by induction on n, that

$$x(nt) = [x(t)]^n \tag{24}$$

for any integer n and any real t such that $|nt| < \delta$. Now for any real t, $\left|\frac{1}{n}t\right| < \delta$ holds for some positive integer n; $\left[x\left(\frac{1}{n}t\right)\right]^n$ is then defined as an element of G and it depends only on t, not on n. For if we have also $\left|\frac{1}{m}t\right| < \delta$, then

$$\left[x\left(\frac{1}{n}t\right)\right]^n = \left[x\left(\frac{1}{mn}t\right)\right]^{mn} = \left[x\left(\frac{1}{m}t\right)\right]^m,$$

by (24). We may therefore put

$$x^*(t) = \left[x\left(\frac{1}{n}t\right)\right]^n, \quad \text{where } n \text{ satisfies } \left|\frac{1}{n}t\right| < \delta. \tag{25}$$

This defines a mapping of R into G which extends the mapping $x(t)$; for when both these mappings are defined then $|t| < \delta$ and we can choose $n = 1$ in (25). Now if $t, t' \in R$, we can find a positive integer n such that $\left|\frac{1}{n}t\right|$, $\left|\frac{1}{n}t'\right|$, $\left|\frac{1}{n}(t+t')\right|$, are all less than δ.

By (23),
$$x\left(\frac{1}{n}t\right)x\left(\frac{1}{n}t'\right) = x\left(\frac{1}{n}(t+t')\right), \tag{26}$$

and $x\left(\frac{1}{n}t'\right)x\left(\frac{1}{n}t\right) = x\left(\frac{1}{n}(t+t')\right)$, which shows that $x\left(\frac{1}{n}t\right)$ and $x\left(\frac{1}{n}t'\right)$ commute. Raising (26) to the nth power, we obtain

$$x^*(t)\,x^*(t') = x^*(t+t');$$

this proves that x^* is a homomorphism and the lemma is established.

Combining this lemma with the remarks following Theorem 2.9.1, we see that for any 1-dimensional Lie group $\mathfrak{G}$ there exists a homomorphism of $\mathfrak{R}$ into $\mathfrak{G}$ which is analytic and invertible at 0. The image of $\mathfrak{R}$ under this homomorphism is a subgroup $\mathfrak{G}_0$ of $\mathfrak{G}$ which is connected (as the continuous image of $\mathfrak{R}$) and contains the nucleus on which the chart $(\bar{x})$ is defined. Therefore $\mathfrak{G}_0$ is the identity component of $\mathfrak{G}$ (Theorem 2.8.3).

Let N be the kernel of this homomorphism. Then N is a closed subgroup of $\mathfrak{R}$, and since the homomorphism is one-one in some nucleus of $\mathfrak{R}$, there is an interval $[-\delta, \delta]$ say $(\delta > 0)$ which contains no element $\neq 0$ of N. It follows that the difference between two elements of N cannot be less than δ, and therefore, if $N \neq 0$, it contains a least positive element, α say. Clearly N contains all multiples of α, and conversely, if $\beta \in N$, then for some n, $0 \leqslant \beta - n\alpha < \alpha$. Since $\beta - n\alpha \in N$, we must have $\beta - n\alpha = 0$ by the definition of α; hence $\beta = n\alpha$ and N consists exactly of the multiples of α. The homomorphism of $\mathfrak{R}$ onto $\mathfrak{G}_0$ can then be defined as an isomorphism of $\mathfrak{R}/N$ onto $\mathfrak{G}_0$. Since this isomorphism is analytic, and is invertible at 0, it is invertible everywhere and so it is an analytic isomorphism. Thus $\mathfrak{G}_0$ is analytically isomorphic to $\mathfrak{R}/N$. Now $\mathfrak{R}/N$, the additive group of real numbers mod α, is analytically isomorphic to the group $\mathfrak{T}$ of real numbers mod 1, an obvious isomorphism being $t \to \frac{1}{\alpha} t$. Altogether we have proved then that $\mathfrak{G}_0$ is analytically isomorphic to $\mathfrak{T}$, unless $N = 0$. In that case we can deduce as before that $\mathfrak{G}_0$ is analytically isomorphic to $\mathfrak{R}$; thus we have proved

THEOREM 2.9.3. *Any connected* 1-*dimensional Lie group is analytically isomorphic either to the additive group of real numbers* $\mathfrak{R}$, *or to* $\mathfrak{T}$, *the additive group of real numbers* mod 1. *In particular, every* 1-*dimensional Lie group is locally isomorphic to* $\mathfrak{R}$.

CHAPTER III

THE LIE ALGEBRA OF A LIE GROUP

3.1. The commutator of two infinitesimal transformations. Let $\mathfrak{M}$ be an analytic manifold; we wish to consider its space $\mathfrak{L}$ of analytic infinitesimal transformations in greater detail. Each element of $\mathfrak{L}$ maps the set $\mathscr{A}$ of all analytic functions in $\mathfrak{M}$ into itself, and we may therefore define, for any $X, Y \in \mathfrak{L}$, a mapping XY by the equation

$$(XY)f = X(Yf) \quad (f \in \mathscr{A}).$$

Clearly XY is again a linear mapping of $\mathscr{A}$ into itself. It does not in general belong to $\mathfrak{L}$, for if in terms of some admissible chart, we have $X = \xi^i . \partial/\partial x^i$, $Y = \eta^i . \partial/\partial x^i$, then

$$XYf = \xi^i \frac{\partial}{\partial x^i}\left(\eta^j \frac{\partial f}{\partial x^j}\right)$$

$$= \xi^i \frac{\partial \eta^j}{\partial x^i}\frac{\partial f}{\partial x^j} + \xi^i \eta^j \frac{\partial^2 f}{\partial x^i \partial x^j}.$$

Owing to the presence of the second derivatives of f this is not an infinitesimal transformation. However, if we calculate YXf, then the coefficients of the second derivatives of f will be the same as in XYf; thus we find that

$$XYf - YXf = \xi^i \frac{\partial \eta^j}{\partial x^i}\frac{\partial f}{\partial x^j} - \eta^i \frac{\partial \xi^j}{\partial x^i}\frac{\partial f}{\partial x^j},$$

and it follows that the operation

$$XY - YX = \left(\xi^i \frac{\partial \eta^j}{\partial x^i} - \eta^i \frac{\partial \xi^j}{\partial x^i}\right)\frac{\partial}{\partial x^j}$$

is an infinitesimal transformation. It is clearly analytic, since X and Y are, and so we have

THEOREM 3.1.1. *If X and Y are in $\mathfrak{L}$, then the operation $[X, Y]$ defined by*

$$[X, Y] = XY - YX \tag{1}$$

again belongs to $\mathfrak{L}$.

The element $[X, Y]$ is called the *commutator* or *Lie product* of X and Y. If we regard this Lie product as a multiplication which is defined on $\mathfrak{L}$, then $\mathfrak{L}$ is a linear algebra† over the field R. More precisely we have

THEOREM 3.1.2. *The Lie multiplication defined on* $\mathfrak{L}$ *by* (1) *is linear in each factor, i.e.*

$$\left.\begin{aligned}[\alpha X+\beta Y, Z]&=\alpha[X, Z]+\beta[Y, Z]\\ [X, \alpha Y+\beta Z]&=\alpha[X, Y]+\beta[X, Z]\end{aligned}\right\} \quad (X, Y, Z \in \mathfrak{L};\ \alpha, \beta \in R), \quad (2)$$

and it satisfies the laws

$$[X, X]=0, \quad (3)$$

$$[X, Y]+[Y, X]=0, \quad (4)$$

$$[[X, Y], Z]+[[Y, Z], X]+[[Z, X], Y]=0. \quad (5)$$

Equation (5) is known as the *Jacobi identity*.

Proof. The equations (3) and (4) are an immediate consequence of (1). To prove (5) we have

$$[[X, Y], Z]=XYZ-YXZ-ZXY+ZYX.$$

Permuting X, Y and Z cyclically and adding the terms so obtained, we reach zero, whence (5) follows.

Any linear algebra satisfying (3) and (5) is called a *Lie algebra*. Such an algebra satisfies (2) by its definition as a linear algebra, and (4) is easily seen to be a consequence of (2) and (3). Thus we may express Theorem 3.1.2 by saying that the space $\mathfrak{L}$ is a Lie algebra with respect to the Lie multiplication.

Let $\mathfrak{M}_1$ and $\mathfrak{M}_2$ be analytic manifolds and write

$$\mathfrak{L}_i=\mathfrak{L}(\mathfrak{M}_i) \quad (i=1, 2).$$

We saw in 1.8 that any analytic homeomorphism Φ of $\mathfrak{M}_1$ onto $\mathfrak{M}_2$ induces a mapping $X \to X^{d\Phi}$ of $\mathfrak{L}_1$ onto $\mathfrak{L}_2$, defined by

$$X^{d\Phi}f=\{\Phi^{-1}\}X\{\Phi\}f. \quad (6)$$

† By a linear algebra over R we understand a vector space V over R (possibly of infinite dimension) with a binary operation xy defined on V with values in V, such that

$$(\alpha x+\beta y)z=\alpha xz+\beta yz, \quad x(\alpha y+\beta z)=\alpha xy+\beta xz \quad (x, y, z \in V,\ \alpha, \beta \in R).$$

We do not assume this operation to be associative.

As an immediate consequence of this definition we have

THEOREM 3.1.3. *Let Φ be an analytic homeomorphism of $\mathfrak{M}_1$ onto $\mathfrak{M}_2$; then $d\Phi$ is a mapping of $\mathfrak{L}_1$ onto $\mathfrak{L}_2$ with inverse $d\Phi^{-1}$ and*

$$\left.\begin{aligned}(\alpha X+\beta Y)^{d\Phi}&=\alpha X^{d\Phi}+\beta Y^{d\Phi}\\ [X, Y]^{d\Phi}&=[X^{d\Phi}, Y^{d\Phi}]\end{aligned}\right\} \quad (X, Y\in\mathfrak{L}_1;\ \alpha,\beta\in R). \qquad \begin{matrix}(7)\\(8)\end{matrix}$$

Equation (7) just states that $d\Phi$ is a linear mapping, and this is easily seen to be the case. To prove (8), we have, for X, $Y\in\mathfrak{L}_1$ and any $f\in\mathscr{A}(\mathfrak{M}_2)$,

$$\begin{aligned}[X^{d\Phi}, Y^{d\Phi}]f&=X^{d\Phi}Y^{d\Phi}f-Y^{d\Phi}X^{d\Phi}f\\ &=\{\Phi^{-1}\}X\{\Phi\}\{\Phi^{-1}\}Y\{\Phi\}f-\{\Phi^{-1}\}Y\{\Phi\}\{\Phi^{-1}\}X\{\Phi\}f\\ &=\{\Phi^{-1}\}XY\{\Phi\}f-\{\Phi^{-1}\}YX\{\Phi\}f\\ &=\{\Phi^{-1}\}(XY-YX)\{\Phi\}f\\ &=[X, Y]^{d\Phi}f,\end{aligned}$$

which was to be proved.

The assertion of this theorem is usually expressed by saying that the mapping $d\Phi$ is an *isomorphism* of the Lie algebra $\mathfrak{L}_1$ onto $\mathfrak{L}_2$. If $\mathfrak{M}_1$ and $\mathfrak{M}_2$ coincide, then so do $\mathfrak{L}_1$ and $\mathfrak{L}_2$ and the theorem states in this case that $d\Phi$ is an isomorphism of $\mathfrak{L}_1$ with itself, i.e. it is an *automorphism* of $\mathfrak{L}_1$.

Ex. 1. If A is any associative algebra, then the underlying vector space of A is a Lie algebra with respect to the multiplication $[x,y]=xy-yx$.

Ex. 2. The vectors in 3-dimensional Euclidean space form a Lie algebra with respect to the usual vector multiplication $x\wedge y$.

3.2. The algebra of infinitesimal right translations.

We now specialize by taking our manifold to be a Lie group $\mathfrak{G}$. Each element a of $\mathfrak{G}$ defines on $\mathfrak{G}$ the right translation

$$\rho_a\colon x\to xa \quad (x\in\mathfrak{G}),$$

and $\mathfrak{G}$ is completely determined by these mappings ρ_a. Our object is to define a set $\mathfrak{g}$ of infinitesimal right translations and to show how $\mathfrak{G}$ may be generated, at least locally, by these operations. We shall find that $\mathfrak{g}$ is a subspace of $\mathfrak{L}$ which is closed under

Lie multiplication and hence forms a Lie algebra; its properties will, to a large extent, reflect the local properties of $\mathfrak{G}$.

We introduce the following notation: If f is a function defined in $\mathfrak{G}$, and $a \in \mathfrak{G}$, we define a function f^a in $\mathfrak{G}$ by the equation

$$f^a(x) = f(ax). \tag{9}$$

If $f \in \mathscr{A}_{ax}$ then $f^a \in \mathscr{A}_x$; further we note the rule

$$(f^a)^b = f^{ab}, \tag{10}$$

which holds because $(f^a)^b\,(x) = f^a(bx) = f(abx) = f^{ab}(x)$.

Definition. Let L be a tangent vector at the unit element e of a Lie group $\mathfrak{G}$; the infinitesimal transformation X defined by

$$X_x f = L f^x \quad (x \in \mathfrak{G};\ f \in \mathscr{A}(\mathfrak{G})), \tag{11}$$

is called the *infinitesimal right translation* defined by L. The set of all infinitesimal right translations of $\mathfrak{G}$ obtained in this way is denoted by $\Lambda(\mathfrak{G})$ or, more briefly, by $\mathfrak{g}$.

To justify the name 'right translation' let us work out an explicit expression for X, defined by (11). Suppose that in terms of some chart at e, L is given by

$$L = \lambda^i \frac{\partial}{\partial x^i},$$

while the multiplication near e is given by the equations

$$(xy)^i = \phi^i(x, y),$$

with analytic composition functions ϕ^i. Let $f \in \mathscr{A}_x$, where x is near e, then $f^x(y) = f(xy)$ is defined for y near e, and

$$X_x f = L f^x = \lambda^i \left[\frac{\partial f(xy)}{\partial y^i}\right]_{y=e}.$$

If we put $xy = z$, this becomes

$$X_x f = \lambda^i \left[\frac{\partial f}{\partial z^k}\frac{\partial \phi^k(x, y)}{\partial y^i}\right]_{y=e}. \tag{12}$$

To see that this represents an 'infinitesimal' right translation, in the intuitive sense, let us put $y^i = e^i + \lambda^i \delta t$ and neglect powers of δt beyond the first. Then the mapping $x \rightarrow xy$ may be regarded as an 'infinitesimal' right translation; the corresponding change in f is

$$\delta f = f(z) - f(x) = \left[\frac{\partial f}{\partial z^k}\frac{\partial \phi^k(x,y)}{\partial y^i}\right]_{y=e} \lambda^i \delta t.$$

A comparison with (12) shows that Xf represents the change in f at x, per unit change of the parameter t, under the right translation in the direction given by λ^i.

In order to bring (12) to a more convenient form, we introduce the following abbreviation:

$$\psi^i_j(x) = \left[\frac{\partial \phi^i(x,y)}{\partial y^j}\right]_{y=e}. \qquad (13)$$

We shall refer to the functions ψ^i_j defined here as the *transformation functions* of the group (in the given chart). In the 1-dimensional case they reduce to the single function which we have already encountered in 2.9.

Using (13), we can write (12) as

$$X_x f = \lambda^i \psi^j_i(x)\frac{\partial f}{\partial x^j}.$$

Hence we obtain as the symbol of the infinitesimal right translation defined by $L = \lambda^i . \partial/\partial x^i$,

$$X = \lambda^i \psi^j_i(x)\frac{\partial}{\partial x^j}. \qquad (14)$$

The first fact we prove about $\mathfrak{g}$ is that it is a vector space isomorphic to $\mathfrak{L}_e$:

THEOREM 3.2.1. *Let $\mathfrak{G}$ be a Lie group, then the set $\mathfrak{g}$ of its infinitesimal right translations is a subspace of $\mathfrak{L}$, the space of all analytic infinitesimal transformations on $\mathfrak{G}$ and, qua vector space over R, $\mathfrak{g}$ is isomorphic to the space $\mathfrak{L}_e$ of tangent vectors at e.*

Proof. That an infinitesimal right translation is analytic at e follows from (14), because the transformation functions ψ^j_i are analytic. Now the equation $X_{ax} f = Lf^{ax} = X_x f^a$ shows that $Xf \in \mathscr{A}_a$ whenever $f \in \mathscr{A}_a$, so that X is analytic at any point a of $\mathfrak{G}$.

Thus the rule (11) associating an infinitesimal right translation with each tangent vector is a mapping of $\mathfrak{L}_e$ into $\mathfrak{L}$. It is immediate from (11) that this mapping is linear and it is one-one, since if L_1 and L_2 both map into X, then $L_1 = X_e = L_2$. Further, the mapping is onto $\mathfrak{g}$ by definition. This completes the proof.

Since $\mathfrak{L}_e$ has the same dimension as the Lie group $\mathfrak{G}$ (Theorem 1.4.2), we have the

COROLLARY. *The dimension of* $\mathfrak{g}$ *is equal to the dimension of* $\mathfrak{G}$.

In an abstract group the right translations may be characterized as the mappings of the group into itself which are left-invariant, i.e. which commute with the left translations. For

$$x\lambda_a \rho_b = (ax)\,b = a(xb) = x\rho_b \lambda_a,$$

and conversely, if the mapping θ satisfies $(ax)^\theta = ax^\theta$, then $a^\theta = ae^\theta$, so that $\theta = \rho_b$, where $b = e^\theta$. There is a similar characterization of infinitesimal right translations in a Lie group. We shall say that an analytic infinitesimal transformation X is *left-invariant*, if

$$X^{d\lambda_a} = X \quad \text{for all } a \in \mathfrak{G}. \tag{15}$$

This definition has a meaning because the left translation $d\lambda_a$ is an analytic homeomorphism of $\mathfrak{G}$; moreover, the condition is analogous to that for abstract groups, as we see if we write (15) in the form

$$\{\lambda_a^{-1}\}\, X\{\lambda_a\} f = Xf \quad (a \in \mathfrak{G}, f \in \mathscr{A}(\mathfrak{G})).$$

Since $\{\lambda_a\} f = f^a$, by (9), the condition for left-invariance may also be written

$$(Xf)^a = Xf^a \quad (a \in \mathfrak{G}, f \in \mathscr{A}(\mathfrak{G})). \tag{16}$$

From this equation it is easy to see that every element of $\mathfrak{g}$ is left-invariant: if $X \in \mathfrak{g}$, then $X_{ax} f = X_e f^{ax} = X_x f^a$, and this expresses the fact that the two sides of (16) agree at x. Conversely, when (16) holds, then $X_{ax} f = X_x f^a$ for all $a, x \in \mathfrak{G}$, whence $X_a f = X_e f^a$, which shows that X is the infinitesimal right translation defined by X_e. This proves

THEOREM 3.2.2. *An analytic infinitesimal transformation on a Lie group is a right translation if and only if it is left-invariant.*

The set $\mathfrak{g}$ of infinitesimal right translations on a Lie group $\mathfrak{G}$ is known as the *Lie algebra* of $\mathfrak{G}$; this terminology is justified by

THEOREM 3.2.3. *The set* $\mathfrak{g}$ *of infinitesimal right translations of a Lie group* $\mathfrak{G}$ *forms a Lie algebra (under Lie multiplication), of the same dimension as* $\mathfrak{G}$.

Proof. We saw already that $\mathfrak{g}$ is a subspace of $\mathfrak{L}$ of dimension equal to the dimension of $\mathfrak{G}$ (Theorem 3.2.1 and Corollary). If $X, Y \in \mathfrak{g}$, then by Theorem 3.1.3, and the left-invariance of X and Y,

$$[X, Y]^{d\lambda_a} = [X^{d\lambda_a}, Y^{d\lambda_a}] = [X, Y].$$

Hence $[X, Y]$ is again left-invariant and so belongs to $\mathfrak{g}$, as we wished to show.

In order to obtain a basis for $\mathfrak{g}$, we take a chart at e. Then the tangent vector $\partial/\partial x^i$ defines the element

$$X_i = \psi_i^j(x)\frac{\partial}{\partial x^j} \tag{17}$$

of $\mathfrak{g}$. Since the $\partial/\partial x^i$ form a basis of $\mathfrak{L}_e$, the corresponding elements X_i of $\mathfrak{g}$ form a basis of $\mathfrak{g}$ (Theorem 3.2.1). They are sometimes called *the infinitesimal transformations of the group* (in the given chart).

Since $\mathfrak{g}$ is a Lie algebra, the elements X_i defined by (17) satisfy equations of the form

$$[X_i, X_j] = c_{ij}^k X_k,$$

where the c_{ij}^k are n^3 real constants, the multiplication constants of the algebra $\mathfrak{g}$. The c_{ij}^k are also known as the *structure constants* of the group $\mathfrak{G}$. These constants satisfy certain relations, namely, those implied by (4) and (5):

$$\left.\begin{aligned} c_{ij}^k + c_{ji}^k &= 0 \qquad (18)\\ c_{ij}^r c_{rk}^s + c_{jk}^r c_{ri}^s + c_{ki}^r c_{rj}^s &= 0 \qquad (19)\end{aligned}\right\} \quad (i, j, k, s = 1, \ldots, n).$$

Conversely, if on an n-dimensional vector space V over R, with the basis $v_1, \ldots, v_n$, we define a multiplication by putting

$$v_i v_j = c_{ij}^k v_k,$$

where the c's are n^3 real constants satisfying (18) and (19), and extend the definition to the whole of V by the linearity conditions (2), we obtain a Lie algebra. It can be shown that any Lie algebra of finite dimension over R, defined in this abstract way, is in fact the Lie algebra of some Lie group.†

† For the proof that a local Lie group can be constructed for any given Lie algebra, see Chapter V.

3.3. Lie groups of transformations. In the definition of the Lie algebra of a group $\mathfrak{G}$, the group played a double role: in the first place it was the underlying analytic manifold and secondly it provided the right translations, which form a group (homeomorphic to $\mathfrak{G}$) consisting of homeomorphisms of the manifold. It is of course possible to separate these two roles; this leads to the notion of a group of transformations of a manifold. That is the way in which Lie groups first arose, and it may be of interest to consider very briefly how the Lie algebra of the group would be defined in this case.

A Lie group $\mathfrak{G}$ is said to be a *group of transformations* of an analytic manifold $\mathfrak{M}$, or, more briefly, to *act* on $\mathfrak{M}$, if to each pair of elements $p \in \mathfrak{M}$, $x \in \mathfrak{G}$ there corresponds an element $q \in \mathfrak{M}$, denoted by px, such that

TR. 1. *The mapping* $(p, x) \to px$ *of* $\mathfrak{M} \times \mathfrak{G}$ *into* $\mathfrak{M}$ *is analytic*,

TR. 2. $pe = p$ *for all* $p \in \mathfrak{M}$,

TR. 3. $(px)y = p(xy)$ *for all* $p \in \mathfrak{M}$, $x, y \in \mathfrak{G}$.

If $x = e$ is the only element of $\mathfrak{G}$ which satisfies $px = p$ for all $p \in \mathfrak{M}$, then the group is said to be *effective*.† It is clear that any Lie group acts effectively on itself by right translations.

Let $(x) = (x^1, \ldots, x^n)$ be a chart at $e \in \mathfrak{G}$, and $(p) = (p^1, \ldots, p^\nu)$ a chart at some point p_0 of $\mathfrak{M}$. Then we can express TR. 1 as

$$q^\alpha = \phi^\alpha(p, x) \quad (\alpha = 1, \ldots, \nu),$$

with analytic functions ϕ^α. If we put

$$\psi_i^\alpha(p) = \left[\frac{\partial \phi^\alpha(p, x)}{\partial x^i}\right]_{x=e},$$

then the space spanned by the infinitesimal transformations

$$X_i = \psi_i^\alpha(p) \frac{\partial}{\partial p^\alpha}$$

is closed under Lie multiplication, and hence is a Lie algebra. If the group is effective, then it can be shown that the X_i are linearly independent (over R) and the resulting Lie algebra is isomorphic to the Lie algebra of $\mathfrak{G}$ defined in 3.2.

† In the older terminology this was expressed by saying that the parameters of this group (i.e. the coordinates of a chart at e) were *essential*.

As an illustration we consider the group of affine transformations on a line:

$$q = x^1 p + x^2 \quad (x^1 > 0).$$

If $y = (1 + \delta x^1, \delta x^2)$ differs by very little from $e = (1, 0)$, then

$$q = p(1 + \delta x^1) + \delta x^2,$$

and

$$f(q) - f(p) = p \frac{df}{dp} \delta x^1 + \frac{df}{dp} \delta x^2.$$

Thus the infinitesimal transformations of the group are

$$S = p \frac{d}{dp} \quad \text{and} \quad T = \frac{d}{dp}.$$

As is easily seen, we have

$$[S, T] = -T. \tag{20}$$

The element $p.d/dp$ is an infinitesimal stretch:

$$p \to p(1 + \delta x^1),$$

and d/dp is an infinitesimal translation: $p \to p + \delta x^2$. Thus (20) may be interpreted as follows: If we first perform a small stretch S, followed by a small translation T, and then perform these operations in the reverse order, the two results will differ by a small translation. Explicitly we have

$$p(1 + \delta x^1) + \delta x^2 - (p + \delta x^2)(1 + \delta x^1) = -\delta x^1 . \delta x^2.$$

Another example is given by the group O_3 of rotations in three dimensions. This may be regarded as a Lie group acting on R^3. Using x, y, z as coordinates in R^3, we can write the infinitesimal rotations about the axes as

$$L = z \frac{\partial}{\partial y} - y \frac{\partial}{\partial z}, \quad M = x \frac{\partial}{\partial z} - z \frac{\partial}{\partial x}, \quad N = y \frac{\partial}{\partial x} - x \frac{\partial}{\partial y}$$

(cf. 1.7). If we represent these infinitesimal rotations by vectors along the axes of rotation (using the right-hand screw convention), then the operations of the Lie algebra of O_3 are precisely the addition and vector multiplication in the usual sense. The multiplication table of our Lie algebra is given by

$$[L, M] = N, \quad [M, N] = L, \quad [N, L] = M.$$

Ex. 1. Show that in the group of affine transformations on a line (discussed above) the multiplication law is given by

$$(xy)^1 = x^1y^1, \quad (xy)^2 = x^2y^1 + y^2.$$

Determine the infinitesimal transformations of the group and verify that the Lie algebra defined by S and T above is isomorphic to the Lie algebra of the group (cf. (20)).

3.4. The Lie algebra of a subgroup. Let $\mathfrak{g}$ be a Lie algebra and $\mathfrak{h}$ a subspace of $\mathfrak{g}$ such that $[X, Y] \in \mathfrak{h}$ for all $X, Y \in \mathfrak{h}$; then $\mathfrak{h}$ is called a *subalgebra* of $\mathfrak{g}$. We consider a Lie group $\mathfrak{G}$ and an analytic subgroup $\mathfrak{H}$ of $\mathfrak{G}$. The subgroup $\mathfrak{H}$ is again a Lie group and hence has a Lie algebra; we shall determine this Lie algebra as a subalgebra of $\mathfrak{G}$.

Denote the Lie algebra of $\mathfrak{G}$ by $\mathfrak{g}$, and define $\mathfrak{h}$ as the set of all $X \in \mathfrak{g}$ such that $X_e f = 0$ for all $f \in \mathscr{A}(\mathfrak{G})$ which vanish on $\mathfrak{H}$. Given any function $f \in \mathscr{A}(\mathfrak{G})$, we denote again by $\bar{f}$ its restriction to $\mathfrak{H}$. Then the definition of $\mathfrak{h}$ reads

$X \in \mathfrak{h}$ if and only if $X_e f = 0$ for all $f \in \mathscr{A}(\mathfrak{G})$ such that $\bar{f} = 0$.

Let $F \in \mathscr{A}_e(\mathfrak{H})$; by choosing a suitable chart $(x^1, \ldots, x^n)$ at e in $\mathfrak{G}$, we may suppose that $(\bar{x}^1, \ldots, \bar{x}^\nu)$ forms a chart at e in $\mathfrak{H}$, and we may then express F as a function of $\bar{x}^1, \ldots, \bar{x}^\nu$, say

$$F = \theta(\bar{x}^1, \ldots, \bar{x}^\nu).$$

The function F may be regarded as the restriction to $\mathfrak{H}$ of a function defined in $\mathfrak{G}$: $F = \bar{f}$, where $f = \theta(x^1, \ldots, x^\nu)$. Here f is not uniquely determined by F, but if $F = \bar{g}$ also, then $\bar{f} - \bar{g} = 0$, hence $X_e f = X_e g$ for all $X \in \mathfrak{h}$. Thus the mapping $F \to X_e f$, for a given $X \in \mathfrak{h}$, is a single-valued mapping of $\mathscr{A}_e(\mathfrak{H})$ into R. We may therefore define, for each $X \in \mathfrak{h}$, a mapping X'_e of $\mathscr{A}_e(\mathfrak{H})$ into R by the rule

$$X'_e F = X_e f, \quad \text{where} \quad \bar{f} = F.$$

This mapping is clearly linear: If $F = \bar{f}$, $G = \bar{g}$ and $\alpha, \beta \in R$, then $X'_e(\alpha F + \beta G) = X_e(\alpha f + \beta g) = \alpha X_e f + \beta X_e g$; moreover, since

$$\overline{fg} = FG,$$

we have $X'_e(FG) = X_e(fg) = X_e f . g + f . X_e g$. This shows that X'_e is actually a tangent vector at e. By left translation we obtain the corresponding element X' of $\Lambda(\mathfrak{H})$, the Lie algebra of $\mathfrak{H}$: If $c \in \mathfrak{H}$, then $X'_c F = X'_e F^c = X_e f^c = X_c f$. Hence we have

$$X'\bar{f} = \overline{Xf}. \tag{21}$$

Since every $F \in \mathscr{A}(\mathfrak{H})$ is of the form $\bar{f}$, this equation defines X' as an element of $\Lambda(\mathfrak{H})$.

We consider the mapping $X \to X'$ of $\mathfrak{h}$ into $\Lambda(\mathfrak{H})$. Clearly this mapping is linear, and it is a one-one mapping, for if $X' = 0$, then $X_e f = 0$ for all $f \in \mathscr{A}(\mathfrak{G})$, whence $X_e = 0$, and so $X = 0$. Further, if $Z \in \Lambda(\mathfrak{H})$, then X_e, defined by $X_e f = Z_e \bar{f}$, belongs to $\mathfrak{L}_e(\mathfrak{G})$ and the element X' obtained from X_e satisfies $X'_e \bar{f} = X_e f = Z_e \bar{f}$, whence $X'_e = Z_e$, $X' = Z$. Thus the mapping $X \to X'$ is a one-one mapping of $\mathfrak{h}$ onto $\Lambda(\mathfrak{H})$, and it is an isomorphism, since

$$\begin{aligned}[X, Y]'\bar{f} = \overline{[X, Y]f} = \overline{XYf} - \overline{YXf} &= X'\overline{Yf} - Y'\overline{Xf} \\ &= X'Y'\bar{f} - Y'X'\bar{f} \\ &= [X', Y']\bar{f}.\end{aligned}$$

This proves

THEOREM 3.4.1. *If $\mathfrak{G}$ is a Lie group and $\mathfrak{H}$ an analytic subgroup of $\mathfrak{G}$, then the Lie algebra of $\mathfrak{H}$ is isomorphic to the set of all $X \in \mathscr{A}(\mathfrak{G})$ such that $X_e f = 0$ for all $f \in \mathscr{A}(\mathfrak{G})$ which vanish on $\mathfrak{H}$.*

Clearly a subset of $\Lambda(\mathfrak{G})$ isomorphic to the algebra $\Lambda(\mathfrak{H})$ is necessarily a sub*algebra* of $\Lambda(\mathfrak{G})$; Theorem 3.4.1 therefore allows us to regard the Lie algebra of $\mathfrak{H}$ as a subalgebra of $\Lambda(\mathfrak{G})$.

Ex. 1. Show that every 1-dimensional subspace of a Lie algebra $\mathfrak{g}$ is a subalgebra of $\mathfrak{g}$.

Ex. 2. Let $X \to X^*$ $(X \in \Lambda(\mathfrak{H}))$ be the inverse of the isomorphism $X \to X'$ defined in the proof of Theorem 3.4.1. If (x) is a chart at e in $\mathfrak{G}$, and $\bar{x}^i$ is the restriction of x^i to $\mathfrak{H}$, show that

$$X^*f = X\bar{x}^i \frac{\partial f}{\partial x^i}.$$

As an illustration of this formula consider $\mathfrak{R}^3$ as a group under vector addition. The Lie algebra $\Lambda(\mathfrak{R}^3)$ has as a basis the infinitesimal translations $\partial/\partial x$, $\partial/\partial y$, $\partial/\partial z$. The set of points given by $x = t$, $y = -t$, $z = 0$, is an analytic subgroup of $\mathfrak{R}^3$; its Lie algebra is spanned by the translation $T = d/dt$. Then $T^* = \partial/\partial x - \partial/\partial y$.

3.5. One-parameter subgroups. An analytic subgroup of a Lie group $\mathfrak{G}$ determines a subalgebra of $\Lambda(\mathfrak{G})$, as we have just seen, and we may ask whether, conversely, each subalgebra of $\Lambda(\mathfrak{G})$ determines an analytic subgroup of $\mathfrak{G}$. The answer is in the affirmative, as we shall prove in Chapter VI, but for the moment we shall only consider the case of one-dimensional subgroups. We may confine our attention to connected subgroups, for if a subalgebra $\mathfrak{h}$ of $\Lambda(\mathfrak{G})$ corresponds to an analytic subgroup $\mathfrak{H}$, then $\mathfrak{h}$ also corresponds to $\mathfrak{H}_0$, the identity component of $\mathfrak{H}$, and $\mathfrak{H}_0$ is again an analytic subgroup of $\mathfrak{G}$. For brevity we call a connected 1-dimensional analytic subgroup of $\mathfrak{G}$ a *1-parameter subgroup* of $\mathfrak{G}$.

If $\mathfrak{H}$ is any 1-parameter subgroup of $\mathfrak{G}$, and $(x^1, \ldots, x^n)$, (t) are charts at e in $\mathfrak{G}$ and $\mathfrak{H}$ respectively, then by Theorem 1.9.1 we may on $\mathfrak{H}$ express the x^i as analytic functions of t:

$$x^i \mid \mathfrak{H} = g^i(t),$$

where the derivatives dg^i/dt do not all vanish at e. We may further suppose that t is a canonical parameter (Theorem 2.9.1), so that

$$g(t)\, g(t') = g(t+t').$$

The infinitesimal transformation of $\mathfrak{H}$ in the chart (t) is then d/dt, and the corresponding element of $\Lambda(\mathfrak{G})$ is X, defined by

$$Xf = \frac{d}{dt}(f \mid \mathfrak{H}) = \frac{dg^i}{dt}\frac{\partial f}{\partial x^i}. \qquad (22)$$

On the other hand, every non-zero element $X = \xi^i . \partial/\partial x^i$ of $\Lambda(\mathfrak{G})$ forms a basis of a 1-dimensional subalgebra, and to find the corresponding analytic subgroups we have to solve the equations

$$\frac{dx^i}{dt} = \xi^i(x),$$

and show that the solution does in fact define a 1-parameter subgroup. This is what we shall do in the proof of the next theorem.

THEOREM 3.5.1. *Let $\mathfrak{G}$ be an Lie group; given any $X \in \Lambda(\mathfrak{G})$, $X \neq 0$, there is a unique 1-parameter subgroup $\mathfrak{H}$ of $\mathfrak{G}$ whose infinitesimal transformation is X.*†

† Although the subgroup $\mathfrak{H}$ is uniquely determined by X, $\mathfrak{H}$ determines only the subalgebra generated by X, so that X itself is only determined to within a non-zero factor (which depends essentially on the choice of the coordinate in $\mathfrak{H}$).

Proof. Take any chart $(x^1, \ldots, x^n)$ at e in $\mathfrak{G}$, and let $\psi^j_i(x)$ be the transformation functions. Then X has the form

$$X = \lambda^i \psi^j_i(x) \frac{\partial}{\partial x^j},$$

for some constants λ^i, not all zero. Now consider the equations

$$\frac{dx^k}{dt} = \lambda^i \psi^k_i(x). \tag{23}$$

These equations (23), together with the initial conditions $x^i = e^i$ when $t = 0$, determine a unique solution

$$x^i = g^i(t), \tag{24}$$

where the g^i are analytic at $t = 0$; since $dg^i/dt = \lambda^i$ when $t = 0$, the derivatives dg^i/dt do not all vanish at $t = 0$. We now show that

$$g(t)\, g(t') = g(t + t'), \tag{25}$$

whenever both t and t' are sufficiently small. For this purpose we write $h(t, t') = g(t)\, g(t')$ and show that $g(t + t')$ and $h(t, t')$ satisfy the same system of differential equations with given initial conditions, namely,

$$\left.\begin{aligned} \frac{\partial z^k}{\partial t'} &= \lambda^i \psi^k_i(z), \\ z^i &= g^i(t) \quad \text{when } t' = 0. \end{aligned}\right\} \tag{26}$$

Here t' is the independent variable, while t is regarded as a parameter. It then follows, by Theorem A. 1 of the Appendix, that $h(t, t') = g(t + t')$, i.e. (25). The initial conditions are clearly satisfied: $g(t + 0) = g(t)$, and $h(t, 0) = g(t)\, g(0) = g(t)\, e = g(t)$. Next we have

$$\frac{\partial g^k(t + t')}{\partial t'} = \lambda^i \psi^k_i(g(t + t')),$$

and so $g(t + t')$ satisfies (26). To show that h satisfies it, we differentiate the equation

$$\phi^i(x, yv) = \phi^i(xy, v)$$

with respect to v^k and then put $v = e$. This gives us

$$\frac{\partial (xy)^i}{\partial y^j} \psi^j_k(y) = \psi^i_k(xy), \tag{27}$$

valid for any x and y near e.

Now let us write $x=g(t)$, $y=g(t')$, and differentiate $h(t,t')=xy$ with respect to t':

$$\frac{\partial h^i(t,t')}{\partial t'}=\frac{\partial(xy)^i}{\partial y^j}\frac{dy^j}{dt'}$$

$$=\frac{\partial(xy)^i}{\partial y^j}\psi^j_k(y)\lambda^k \quad \text{by the definition of } y,$$

$$=\psi^i_k(xy)\lambda^k \quad \text{by (27);}$$

this shows that $h(t,t')$ ($=xy$) satisfies (26). Therefore (25) holds for small t and t', and it shows that the elements $g(t)$ of $\mathfrak{G}$ (for all t in some neighbourhood of 0) form a local Lie group. The subgroup $\mathfrak{H}$ generated by this local group is analytic and connected (Theorem 2.8.2) and hence is a 1-parameter subgroup of $\mathfrak{G}$. Its infinitesimal transformation is d/dt and this corresponds to X in $\Lambda(\mathfrak{G})$, as we saw already in (22). Finally, the subgroup $\mathfrak{H}$ is unique, for if $\mathfrak{H}_1$ is any 1-parameter subgroup with infinitesimal transformation X, then when we express the coordinates in $\mathfrak{G}$ as functions of the canonical coordinate in $\mathfrak{H}_1$, they must, for a suitable choice of the constant factor, satisfy (23). Hence the functions are given by (24), so that $\mathfrak{H}_1$ coincides near e with $\mathfrak{H}$. But then it coincides with $\mathfrak{H}$ everywhere, and this establishes the theorem.

For later applications we note the following theorem, which is an almost immediate corollary. By a *local analytic homomorphism* of $\mathfrak{G}$ into $\mathfrak{G}'$ we shall understand an analytic mapping $x\to x'$ of a nucleus U of $\mathfrak{G}$ into $\mathfrak{G}'$ such that $(xy)'=x'y'$ whenever $x, y, xy\in U$.

THEOREM 3.5.2. *Let $\mathfrak{G}$ be a Lie group with the chart $(x^1,\ldots,x^n)$ at e and the transformation functions $\psi^j_i(x)$.*

(i) *If $g^k(t)$ is a solution of the equations*

$$\frac{dx^k}{dt}=\lambda^i\psi^k_i(x) \quad (x^k=0 \text{ when } t=0), \tag{28}$$

then the mapping $t\to(g^1(t),\ldots,g^n(t))$ is a local analytic homomorphism of $\mathfrak{R}$ into $\mathfrak{G}$.

(ii) *Conversely, if the functions $g^k(t)$ are such that the mapping $t \to (g^1(t), \ldots, g^n(t))$ is a local analytic homomorphism of $\mathfrak{R}$ into $\mathfrak{G}$, then the $g^k(t)$ satisfy the equations* (28) *with* $\lambda^i = \left[\dfrac{dg^i(t)}{dt}\right]_{t=0}$.

The first part follows from the proof of Theorem 3.5.1 (with $g^i(t) = 0$, in case $\lambda^i = 0$, $i = 1, \ldots, n$). To prove (ii) we note that under the hypothesis the image points form a local analytic subgroup of $\mathfrak{G}$. Let $\mathfrak{H}$ be the subgroup of $\mathfrak{G}$ generated by this local group. Then by Theorem 2.8.2, $\mathfrak{H}$ is an analytic subgroup of $\mathfrak{G}$, and since $g(t)\,g(t') = g(t+t')$ by hypothesis, we infer from Lemma 2.9.2 that $\mathfrak{H}$ is a homomorphic image of $\mathfrak{R}$. Thus either $\mathfrak{H}$ is a 1-parameter subgroup of $\mathfrak{G}$ and the theorem follows from Theorem 3.5.1 or $\mathfrak{H} = \{e\}$, in which case $\lambda^i = 0$ and the theorem is trivially true.

A physical interpretation of the situation in Theorem 3.5.1 may be obtained as follows:† We may think of the group $\mathfrak{G}$ as an n-dimensional space filled with a compressible fluid moving with a velocity whose components at the point x are $\xi^k(x)$ (stationary flow). In the above case $\xi^k(x) = \lambda^i \psi_i^k(x)$, so that the $\xi^k(x)$ are nowhere all zero. Thus there are no points where the fluid is at rest.

If f is any function of position—e.g. the temperature—which is independent of the time, then $\partial f / \partial t = 0$. But the value of f changes for an observer moving with the fluid. The change is

$$\xi^i \frac{\partial f}{\partial x^i} = Xf$$

(differentiation following the motion). As we follow a particle of the fluid it traces out a curve which is uniquely determined by any one point on it. In Theorem 3.5.1 we started at e and obtained the 1-parameter subgroup $\mathfrak{H}$ whose infinitesimal transformation was X. If instead we start at a general point a of $\mathfrak{G}$, we obtain the set of points reached from a by right translation along $\mathfrak{H}$, i.e. the left coset $a\mathfrak{H}$.

The analogue of Theorem 3.5.1 for transformation groups states that to any infinitesimal transformation X on an analytic manifold $\mathfrak{M}$ there corresponds a 1-dimensional Lie group acting

† Cf. the Introduction.

on $\mathfrak{M}$ with infinitesimal transformation X. The proof is similar to that of Theorem 3.5.1; we omit it and instead give an illustration. We take the manifold to be $\mathfrak{R}^2$ and the infinitesimal transformation $X = x\frac{\partial}{\partial x} + y\frac{\partial}{\partial y}$. To obtain a 1-dimensional group of transformations with infinitesimal transformation X, we write, for any function f on $\mathfrak{R}^2$,

$$\frac{df}{dt} = x\frac{\partial f}{\partial x} + y\frac{\partial f}{\partial y}.$$

In particular, $dx/dt = x$, $dy/dt = y$, whence $x = x_0 \exp t$, $y = y_0 \exp t$ and we obtain a 1-dimensional group.

An alternative method of obtaining the group will be to ask for solutions of $\partial f/\partial x + \partial f/\partial y = 0$, i.e. functions which are invariant under the group operations. These are just the homogeneous functions of degree 0, by Euler's theorem; and in fact these functions are invariant under the change $x \to x \exp t$, $y \to y \exp t$.

We may interpret $Xf = x\frac{\partial f}{\partial x} + y\frac{\partial f}{\partial y}$ as the rate of change of f under uniform stretching of the plane. Thus, regarded as an infinitesimal transformation, X is an infinitesimal stretch, and the 1-dimensional group of transformations corresponding to it is obtained by an iteration of this process.

3.6. Taylor's theorem for Lie groups. In the interpretation of Theorem 3.5.1 given in the last section we saw that a 1-dimensional subgroup is obtained by iterating its infinitesimal transformation. We shall now make this idea more precise by writing down an explicit solution of the equations (23).

Let $\mathfrak{G}$ be a Lie group and $X \neq 0$ an element of $\Lambda(\mathfrak{G})$. To X corresponds, by Theorem 3.5.1, a 1-parameter subgroup $\mathfrak{H}$:

$$x^i = g^i(t), \quad g(0) = e,$$

such that for any function $f \in \mathscr{A}(\mathfrak{G})$,

$$\dot{f}(g(t)) = Xf.$$

This equation states how f changes for an infinitesimal displacement along $\mathfrak{H}$. Let us calculate the change for a finite displacement. For brevity we write

$$f(g^1(t), \ldots, g^n(t)) = F(t).$$

Then F is analytic, and by Taylor's theorem,

$$F(t) = F(0) + tF'(0) + \frac{t^2}{2!}F''(0) + \dots$$
$$= f(e) + t[Xf]_e + \frac{t^2}{2!}[X^2 f]_e + \dots$$

for all sufficiently small values of t. Thus

$$f(g(t)) = \left[\left(1 + tX + \frac{t^2X^2}{2!} + \dots\right)f\right]_e,$$

or, more concisely, $f(g(t)) = [\exp(tX)f]_e$, (29)

where $$\exp(tX) = 1 + tX + \frac{t^2X^2}{2!} + \dots.$$

Equation (29) may be regarded as Taylor's theorem for Lie groups; the values of f on the 1-parameter subgroup $\mathfrak{H}$ are given in terms of the derivatives of f at the unit element e. The familiar form of Taylor's theorem for real variables is just the special case where $\mathfrak{G} = \mathfrak{H} = \mathfrak{R}$. Then $X = d/dx$ and

$$f(t) = \sum_{n=0}^{\infty} \frac{t^n}{n!}\left[\frac{d^n f}{dx^n}\right]_{x=0} = [\exp(tX)f]_{x=0}.$$

At any point x of $\mathfrak{R}$ we may write $f(x+t) = \exp(tX)f$; here $x \to x+t$ is the right translation in $\mathfrak{R}$ and d/dx is the corresponding infinitesimal transformation. Similarly we may generalize (29) to

$$f(xg(t)) = \exp(tX)f, \tag{30}$$

either by observing that tX corresponds to a right translation, which is now applied at x instead of e, or by applying the left translation λ_x to (29) and using the fact that X is left-invariant.

Equation (29) is enough to define $\mathfrak{H}$, at least locally; for if we take $f = x^k$ in (29), we find

$$g^k(t) = [\exp(tX)x^k]_e. \tag{31}$$

If $X = \xi^i(x)\dfrac{\partial}{\partial x^i}$, say, this means

$$g^k(t) = t\xi^k(e) + \frac{t^2}{2!}[X\xi^k]_e + \dots,$$

where the series on the right is convergent for all sufficiently small values of t.

3.7. The exponential mapping. Let X again be an element of $\Lambda(\mathfrak{G})$ ($X \neq 0$), and $g(t)$ the general point on the corresponding 1-parameter subgroup, as defined by (31). The right translation $\rho_{g(t)}$ is an analytic mapping of $\mathfrak{G}$ onto itself, and therefore induces a mapping of $\mathscr{A}(\mathfrak{G})$ into itself which we denoted by $\{\rho_{g(t)}\}$ in 1.8:

$$\{\rho_{g(t)}\}f(x) = f(xg(t)).$$

By (30) we also have $f(xg(t)) = \exp(tX)f$, and so we obtain

$$\{\rho_{g(t)}\}f = \exp(tX)f; \tag{32}$$

thus $\rho_{g(t)}$ and $\exp(tX)$, qua mappings of $\mathscr{A}(\mathfrak{G})$, have the same effect. We also saw that $g(t)$ is completely determined by tX, for small t explicitly by (31) and for general t by Theorem 3.5.1. So we shall agree to write $\exp(tX)$ for $g(t)$, i.e. we denote the point $g(t)$ by $\exp(tX)$. The expression $\exp(tX)$, which in the first place was a mapping of $\mathscr{A}(\mathfrak{G})$ into itself, has now been identified with an element of $\mathfrak{G}$. It is easily verified from the proof of Theorem 3.5.1, that when X is replaced by λX, $g(t)$ is replaced by $g(\lambda t)$. We may therefore absorb t into X and just consider the mapping $X \to \exp X$ of $\Lambda(\mathfrak{G})$ into $\mathfrak{G}$. It is called the *exponential mapping*.

Since $\exp X$ is an element of a group, we have a multiplication defined between the elements $\exp X$, and we have to consider its significance. Let $X, Y \in \Lambda(\mathfrak{G})$, and write $\exp X = x$, $\exp Y = y$ for short. By (32) and the contravariant character of ρ_x, we have

$$\begin{aligned}\exp X.(\exp Y.f) &= \{\rho_x\}\{\rho_y\}f\\ &= \{\rho_{xy}\}f,\end{aligned}$$

so that we may define $\exp X \exp Y$ as $\{\rho_{xy}\}$. When we introduce a parameter t, we see that the product $\exp(tX).\exp(tY)$ defined in this way represents (for sufficiently small t) the operator obtained by treating $\exp(tX)$ and $\exp(tY)$ as power series in t and multiplying them together.

Thus, for example, if $[X, Y] = 0$, then

$$\exp X.\exp Y = \exp(X+Y),$$

by the multiplication of power series, and so

$$\exp X.\exp Y.f = \exp(X+Y).f \quad \text{if} \quad [X, Y] = 0.\dagger$$

† Strictly speaking we have to introduce a parameter t and prove this result for small t, but we can then show that it holds for all t, as in Lemma 2.9.2.

In particular, taking $Y = -X$, we have

$$\exp X . \exp(-X) . f = \exp 0 . f = f,$$

and therefore
$$\exp(-X) = (\exp X)^{-1}.$$

In general, $\exp X . \exp Y$ cannot be expressed in the form $\exp Z$. However, as we shall see in Chapter VI, it can be so expressed whenever X and Y are sufficiently close to 0, in terms of a certain obvious topology on $\Lambda(\mathfrak{G})$.

Ex. (Cartan). Let G be the multiplicative group of real 2×2 matrices with determinant 1. This becomes a Lie group $\mathfrak{G}$ when any three of the four coefficients are taken as coordinates at e. Show that matrices whose eigenvalues are distinct, real and negative, do not lie on any 1-parameter subgroup of $\mathfrak{G}$, and hence are not of the form $\exp X$ ($X \in \Lambda(\mathfrak{G})$).

CHAPTER IV

THE ALGEBRA OF DIFFERENTIAL FORMS

4.1. The exterior algebra of a vector space. Let us consider for a moment the vectors in ordinary 3-dimensional space. In the well-known graphical representation each vector u is represented by a directed line segment starting at the origin. Similarly, the vector product $u \wedge v$ of two vectors is represented by a certain line segment at right angles to the plane of u and v, and it is therefore identified with the vector represented by this line segment. We shall not make this identification, but regard $u \wedge v$ as a new entity called a *bivector*. The bivectors can be used to describe oriented areas—i.e. plane areas with a sense of going along the boundary—in the same way as vectors serve to describe directed lengths. We note that the orientation of $u \wedge v$ is determined by the order of u and v in the product, thus

$$v \wedge u = -u \wedge v.$$

Analytically $u \wedge v$ is a skew-symmetric tensor of rank 2, its components being the second order minors of the matrix

$$\begin{pmatrix} u_1 & u_2 & u_3 \\ v_1 & v_2 & v_3 \end{pmatrix}$$

formed by the components of u and v.

In a similar way volumes may be described by the triple scalar product of vectors. In elementary vector algebra this is written as $(u \wedge v).w$ or $u.(v \wedge w)$. We shall denote the triple product formed from the vectors u, v, w by $u \wedge v \wedge w$. This triple product is sometimes called a *trivector*; it is again a scalar, or, more accurately, a scalar density, since under linear transformations it is multiplied by the determinant of the transformation. Its unique component is the determinant formed from the components of the vectors u, v, w. Higher products do not occur, since in three dimensions a skew-symmetric tensor of rank greater than three is necessarily zero.

Now let V be any vector space of dimension n over R. There are subspaces of dimension k in V for $0 \leqslant k \leqslant n$, and by an obvious extension of the previous method a k-dimensional volume in V may be described by a product with k factors. This suggests the following construction:

We take a fixed integer k such that $0 \leqslant k \leqslant n$; for each set of integers $i_1, \ldots, i_k$ satisfying $1 \leqslant i_1 < i_2 < \ldots < i_k \leqslant n$ we choose a symbol $a_{i_1 \ldots i_k}$ and denote by V_k the vector space over R on these elements $a_{i_1 \ldots i_k}$ as basis. The dimension of V_k is therefore equal to $\binom{n}{k}$. In particular, for $k = 0$ we obtain a 1-dimensional space V_0, and for $k = 1$ we obtain V_1, a space of dimension n with $a_1, \ldots, a_n$ as a basis.

Now let $j_1, \ldots, j_k$ be any sequence of integers between 1 and n. In case the j's are all distinct there is a unique permutation σ of these k integers such that $j_1\sigma, \ldots, j_k\sigma$ are in ascending order. We put

$$a_{j_1 \ldots j_k} = \begin{cases} a_{j_1\sigma \ldots j_k\sigma} & \text{if } \sigma \text{ is an even permutation,} \\ -a_{j_1\sigma \ldots j_k\sigma} & \text{if } \sigma \text{ is an odd permutation,} \\ 0 & \text{if two of } j_1, \ldots, j_k \text{ are equal.} \end{cases}$$

This defines $a_{j_1 \ldots j_k}$ for any values of the suffixes as an element of V_k, in terms of the given basis of V_k. Next we form the direct sum

$$E = V_0 \oplus V_1 \oplus \ldots \oplus V_n,$$

then E is a vector space with the basis

$$a_{i_1 i_2 \ldots i_k} \quad (i_1 < i_2 < \ldots < i_k;\ k = 0, 1, \ldots, n),$$

and its dimension is therefore $\Sigma \binom{n}{k} = 2^n$. On E we define a multiplication by the rule

$$a_{i_1 \ldots i_r} \wedge a_{j_1 \ldots j_s} = a_{i_1 \ldots i_r j_1 \ldots j_s} \quad (i_1 < \ldots < i_r;\ j_1 < \ldots < j_s). \tag{1}$$

The right-hand side has already been defined when $r + s \leqslant n$; if this does not hold, at least two of $i_1, \ldots, i_r, j_1, \ldots, j_s$ must be equal, and we then take the right-hand side of (1) to be zero. Equation (1) defines the products on the basis elements of E and therefore, by linearity, on the whole of E. In particular, if we permute $i_1, \ldots, i_r$, then both sides of (1) are multiplied by $+1$ or by -1, according as the permutation was even or odd; similarly for the j's. Therefore (1) still holds when the suffixes are in any order, not necessarily ascending. If two of the i's are equal, or two of the j's, then both sides of (1) vanish, so (1) holds in fact for any choice of suffixes.

Now it is easy to verify that the multiplication (1) is associative, and therefore E is a linear associative algebra. For if we write $h=(h_1, \ldots, h_q)$, $i=(i_1, \ldots, i_r)$, $j=(j_1, \ldots, j_s)$ for short, then

$$(a_h \wedge a_i) \wedge a_j = a_{hij} = a_h \wedge (a_i \wedge a_j),$$

and it follows by linearity that the associative law holds for the whole of E. The single basis element a of V_0 is the unit element of E, and the basis $a_1, \ldots, a_n$ of V_1, together with a, forms a set of generators of E, since every element of E is a linear combination of the elements

$$a_{i_1 \ldots i_k} = a_{i_1} \wedge \ldots \wedge a_{i_k}.$$

We shall identify V with the subspace V_1 of E and call E the *exterior algebra* on the vector space V, while the multiplication in E is called *exterior multiplication*. The elements of V_k are said to be homogeneous of degree k, or k-linear, in the elements of V; the monomials of degree k are also called k-vectors. In particular, for $k=1, 2, 3$ we find the vectors, bivectors and trivectors which we met already in the intuitive picture given at the beginning of this section.

Having defined the exterior algebra of a vector space by an explicit construction, we now show that it can also be defined in another, more usual, way, which is particularly convenient for later applications.

THEOREM 4.1.1. *Let V be a vector space and E its exterior algebra. Then E is isomorphic to the associative algebra with a unit element, on a basis of V as generators*† *and with the defining relations*

$$v \wedge v = 0 \quad (v \in V). \tag{2}$$

If V is of dimension n, the dimension of E is 2^n.

Proof. Let $v_1, \ldots, v_n$ be a basis of V, and denote by F the associative algebra with unit element, which is generated by the

† We say that an algebra A with unit element 1 is generated by elements $x_1, x_2, \ldots$ if the least subalgebra of A containing 1 and all the x's is A itself.

elements $v_1, \ldots, v_n$ with the defining relations (2). Since E, the exterior algebra of V, is an associative algebra with unit element, generated by $v_1, \ldots, v_n$ and satisfies (2), it must be a homomorphic image of F, say $E \simeq F/K$, where K, the kernel of the homomorphism, is a subspace of F. It follows that

$$\dim E = \dim F - \dim K;$$

if we can prove that

$$\dim F \leqslant \dim E, \tag{3}$$

then $\dim K$ must be 0. This will imply that $K = 0$ and so E and F will be isomorphic, as asserted by the theorem. To establish (3), we denote by F_k the subspace of F spanned by the products of just k factors in V. Applying (2), we find that

$$v_i \wedge v_i = 0,\dagger \tag{4}$$

and $v_i \wedge v_j + v_j \wedge v_i = (v_i + v_j) \wedge (v_i + v_j) - v_i \wedge v_i - v_j \wedge v_j = 0,\dagger$

whence

$$v_j \wedge v_i = -v_i \wedge v_j. \tag{5}$$

By (5) we can allow for the interchange of adjacent factors in any monomial

$$v_{i_1} \wedge v_{i_2} \wedge \ldots \wedge v_{i_k} \tag{6}$$

simply by changing the sign, and we can therefore bring the suffixes into ascending order by at most changing the sign. The result will be zero if two suffixes are the same, by (4), and so we can express every element of F_k as a linear combination of monomials (6) in which the suffixes are strictly ascending:

$$i_1 < i_2 < \ldots < i_k.$$

There are just $\binom{n}{k}$ of these, so the dimension of F is at most $\Sigma \binom{n}{k} = 2^n$. We saw already that the dimension of E is just 2^n, thus (3) is established and the theorem follows.

4.2. The algebra of differential forms. In Chapter I we defined differentials at a point p of an analytic manifold $\mathfrak{M}$ and saw that they form a space $\mathfrak{L}_p^*$ of dimension n, where n is the dimension of $\mathfrak{M}$ at p. Further, a differential form was defined

† No summation.

as a collection of differentials, one at each point of $\mathfrak{M}$. Our present object is to generalize this concept.

We consider, at each point p of $\mathfrak{M}$, the exterior algebra $\mathfrak{E}(\mathfrak{L}_p^*)$ of the space $\mathfrak{L}_p^*$ and define a *differential form* as a collection of elements of $\mathfrak{E}(\mathfrak{L}_p^*)$, one for each $p \in \mathfrak{M}$. In terms of a chart in $\mathfrak{M}$, a homogeneous form of degree k may be written as

$$\omega = \alpha_{i_1 \dots i_k}(x)\, dx^{i_1} \wedge \dots \wedge dx^{i_k}.$$

The general form is a sum of such homogeneous forms for $k = 0, 1, \dots, n$, but we shall only have to consider homogeneous forms. In fact we only require forms of degree 0, 1 or 2, but it is often simpler to state theorems for the general case.

The forms of degree 0 are just the real functions on $\mathfrak{M}$. Forms of degree 1—i.e. linear forms—are the differential forms as previously defined (cf. 1.7). They are sometimes called *Pfaffian forms* on $\mathfrak{M}$.

The linear form $\alpha_i(x)\, dx^i$ may be interpreted as an element of directed length, for example, in forming the line integral

$$\int \alpha_i(x)\, dx^i$$

taken along some curve in $\mathfrak{M}$. Similarly, the bilinear form

$$\alpha_{ij}(x)\, dx^i \wedge dx^j$$

may be regarded as an element of oriented area.† As an illustration we shall consider the elementary definition of a surface integral.

Given a surface $z = f(x, y)$ in 3-dimensional space, we may wish to integrate f over a region A in the (x, y)-plane. In order to do this we take the element of area at (x, y): $dx \wedge dy$, multiply by $f(x, y)$ and integrate. Thus the required integral is written as

$$\int_A f(x, y)\, dx \wedge dy.$$

With this notation a change of variables takes on a specially simple form: Let $x = x(u, v)$, $y = y(u, v)$, where the functions on the right are continuously differentiable, then

$$dx = x_u du + x_v dv, \quad dy = y_u du + y_v dv,$$

† Strictly speaking, a sum of such elements.

where $x_u = \partial x/\partial u$, etc. Multiplying and observing the rules of exterior multiplication, we find

$$dx \wedge dy = (x_u du + x_v dv) \wedge (y_u du + y_v dv)$$
$$= (x_u y_v - x_v y_u)\, du \wedge dv,$$

thus $$\int f(x,y)\, dx \wedge dy = \int f(x(u,v), y(u,v)) \frac{\partial(x,y)}{\partial(u,v)} du \wedge dv,$$

which is the familiar rule for changing the variables in a double integral.

In an analytic manifold a change of variables near a point p is the passage from one admissible chart at p to another. If the equations of transformation are

$$x^i = \Phi^i(y),$$

then the form $$\alpha_{i_1 \dots i_k} dx^{i_1} \wedge \dots \wedge dx^{i_k}, \tag{7}$$

when expressed in terms of the y's, becomes

$$\alpha_{i_1 \dots i_k} \frac{\partial \Phi^{i_1}}{\partial y^{j_1}} \dots \frac{\partial \Phi^{i_k}}{\partial y^{j_k}} dy^{j_1} \wedge \dots \wedge dy^{j_k}. \tag{8}$$

If the coefficients in (7) are analytic functions at p, then so are the coefficients in (8), since the functions Φ^i are analytic at p. The converse follows by means of the equations of transformation from (x) to (y), and we may therefore define: The differential form ω is called *analytic* at a point p of $\mathfrak{M}$, if in an admissible chart (x) at p it can be written as†

$$\omega = \alpha_{i_1 \dots i_k} dx^{i_1} \wedge \dots \wedge dx^{i_k}, \tag{9}$$

where the α's are analytic at p. By the preceding remark this definition does not depend on the particular chart used. A differential form which is analytic at every point of $\mathfrak{M}$ is simply called *analytic*; for linear forms this definition agrees with the one already given in 1.7. It is clear that the product of two analytic forms is again analytic; hence the analytic differential

† The coefficients in (9) are not uniquely determined by ω, unless we place some restriction on the range of summation, or require the α's to be skew-symmetric in all suffixes. But this does not affect the definition.

forms on $\mathfrak{M}$ constitute a linear algebra, $\mathfrak{C} = \mathfrak{C}(\mathfrak{M})$; we denote by $\mathfrak{C}_k = \mathfrak{C}_k(\mathfrak{M})$ its subspace of homogeneous forms of degree k.

4.3. Exterior differentiation. The elements of $\mathfrak{C}_0 = \mathfrak{C}_0(\mathfrak{M})$, i.e. the analytic differential forms of degree 0, are just the analytic functions on $\mathfrak{M}$. With every such function f we associated in Chapter I the differential form df, defined by the rule

$$\langle L, df \rangle = Lf,$$

where L is any tangent vector. In terms of a chart the definition reads

$$df = \frac{\partial f}{\partial x^i} dx^i.$$

Here d appears as an operator which maps $\mathfrak{C}_0$ into $\mathfrak{C}_1$. We may define d as an operator on $\mathfrak{C}_k$ for general k by the rule:

If $\omega = \alpha_{i_1 \dots i_k} dx^{i_1} \wedge \dots \wedge dx^{i_k}$, then

$$d\omega = d\alpha_{i_1 \dots i_k} \wedge dx^{i_1} \wedge \dots \wedge dx^{i_k}$$
$$= \frac{\partial \alpha_{i_1 \dots i_k}}{\partial x^j} dx^j \wedge dx^{i_1} \wedge \dots \wedge dx^{i_k};$$

by linearity d may then be extended to the whole of $\mathfrak{C}$. This operation is called *exterior differentiation*. For $k = 0, 1$ it reduces to the operations grad and curl which are familiar from vector analysis. We collect the properties of d in

THEOREM 4.3.1. (i) *d is a linear mapping*:

$$d(\lambda\omega + \mu\chi) = \lambda d\omega + \mu d\chi \quad (\omega, \chi \in \mathfrak{C}; \lambda, \mu \in R),$$

(ii) *for any $\omega \in \mathfrak{C}$, $d^2\omega = 0$*,

(iii) *if ω, $\chi \in \mathfrak{C}$ and ω is homogeneous of degree k, then*

$$d(\omega \wedge \chi) = d\omega \wedge \chi + (-1)^k \omega \wedge d\chi.$$

Proof. (i) is clear; to prove (ii), let $\omega = \alpha_{i_1 \dots i_k} dx^{i_1} \wedge \dots \wedge dx^{i_k}$, then $d^2\omega = \dfrac{\partial^2 \alpha_{i_1 \dots i_k}}{\partial x^r \partial x^s} dx^r \wedge dx^s \wedge dx^{i_1} \wedge \dots \wedge dx^{i_k}$. If we interchange r and s then the right-hand side changes sign, hence

$$d^2\omega = -d^2\omega,$$

and so $d^2\omega = 0$. The verification of (iii) is straightforward and is left to the reader.

We note without proof that the properties (i)–(iii) of this theorem, together with the definition of df for $f \in \mathfrak{C}_0$, suffice to define d on the whole of $\mathfrak{C}$.

As we have defined the operation d in terms of a particular chart (at least for forms of positive degree) we still have to show that its effect is independent of this choice. It is convenient to deduce this from the more general fact that the operator d is invariant under any analytic mapping between manifolds. But before proving this we must establish the invariance of the exterior multiplication itself.

Let $\mathfrak{M}$ and $\mathfrak{N}$ be analytic manifolds, and Φ an analytic mapping of $\mathfrak{M}$ into $\mathfrak{N}$. With Φ is associated the mapping $\{\Phi\}$ of $\mathscr{A}(\mathfrak{N})$ into $\mathscr{A}(\mathfrak{M})$; thus $\{\Phi\}$ may also be regarded as a mapping of $\mathfrak{C}_0(\mathfrak{N})$ into $\mathfrak{C}_0(\mathfrak{M})$, and we extend $\{\Phi\}$ to a mapping of $\mathfrak{C}_1(\mathfrak{N})$ into $\mathfrak{C}_1(\mathfrak{M})$ as follows: Write

$$\{\Phi\}(df) = d(\{\Phi\}f); \tag{10}$$

this equation, together with the linearity of $\{\Phi\}$, defines $\{\Phi\}$ on the whole of $\mathfrak{C}_1(\mathfrak{N})$. Explicitly, if (x) and (y) are charts at $p \in \mathfrak{M}$ and $p^\Phi \in \mathfrak{N}$ respectively, and Φ is given by

$$y^i = \Phi^i(x), \tag{11}$$

then
$$\begin{aligned}\{\Phi\}(df) &= \{\Phi\}\left(\frac{\partial f}{\partial y^i}\, dy^i\right)\\ &= \frac{\partial f}{\partial y^i}\frac{\partial \Phi^i}{\partial x^j}\, dx^j.\end{aligned}$$

Generally, we have
$$\{\Phi\}(\alpha_i\, dy^i) = \alpha_i \frac{\partial \Phi^i}{\partial x^j}\, dx^j.$$

Now we extend $\{\Phi\}$ to a mapping of $\mathfrak{C}(\mathfrak{N})$ into $\mathfrak{C}(\mathfrak{M})$ by defining it as a homomorphism on $\mathfrak{C}(\mathfrak{N})$:

$$\{\Phi\}(\omega \wedge \chi) = \{\Phi\}\omega \wedge \{\Phi\}\chi \quad (\omega, \chi \in \mathfrak{C}(\mathfrak{N})).$$

To show that the mapping $\{\Phi\}$ so defined is single-valued on $\mathfrak{C}(\mathfrak{N})$ we have to verify that it preserves the relations in $\mathfrak{C}(\mathfrak{N})$. But any such relation is a consequence of relations of the form

$$\omega \wedge \omega = 0 \quad (\omega \in \mathfrak{C}_1(\mathfrak{N})).$$

When we apply $\{\Phi\}$ to the left-hand side, we obtain $\{\Phi\}\omega \wedge \{\Phi\}\omega$ which vanishes, since $\{\Phi\}\omega \in \mathfrak{C}_1(\mathfrak{M})$. Thus every analytic mapping of $\mathfrak{M}$ into $\mathfrak{N}$ induces a homomorphism of $\mathfrak{C}(\mathfrak{N})$ into $\mathfrak{C}(\mathfrak{M})$. This may be expressed by saying that the mapping $\{\Phi\}$ preserves exterior multiplication. We can now prove the corresponding fact for exterior differentiation:

LEMMA 4.3.2. *Let Φ be an analytic mapping of $\mathfrak{M}$ into $\mathfrak{N}$. Then*

$$\{\Phi\}d\omega = d(\{\Phi\}\omega) \quad \textit{for all } \omega \in \mathfrak{C}(\mathfrak{N}). \tag{12}$$

Proof. We suppose again that Φ is given locally by the equations (11), and for brevity write ω^* instead of $\{\Phi\}\omega$. Since both sides of (12) are linear in ω, we may take ω to be homogeneous of degree k; we shall prove the lemma by induction on k. When $k=1$, we have $\omega = \alpha_i dy^i$, say, and

$$(d\omega)^* = (d\alpha_i \wedge dy^i)^* = (d\alpha_i)^* \wedge (dy^i)^* = d\alpha_i^* \wedge dy^{i*} = d(\alpha_i^* dy^{i*}) = d\omega^*.$$

Next let $k>1$ and assume (12) for forms of degree less than k. Any form of degree k can be written as the sum of forms $\theta \wedge \chi$, where θ and χ have degrees 1 and $k-1$ respectively. Then, using induction and the fact that * is a homomorphism, we have

$$\begin{aligned}(d(\theta \wedge \chi))^* &= (d\theta \wedge \chi - \theta \wedge d\chi)^* \\ &= (d\theta)^* \wedge \chi^* - \theta^* \wedge (d\chi)^* \\ &= d\theta^* \wedge \chi^* - \theta^* \wedge d\chi^* \\ &= d(\theta^* \wedge \chi^*) \\ &= d(\theta \wedge \chi)^*,\end{aligned}$$

and this proves the lemma.

In particular, if $\mathfrak{M} = \mathfrak{N}$ and Φ is an analytic homeomorphism of $\mathfrak{M}$, then we may regard Φ as a transformation of coordinates in $\mathfrak{M}$ (possibly defined only locally). In this case we deduce that the operations of exterior multiplication and differentiation are independent of the particular chart used.

Ex. 1. The homomorphism $\{\Phi\}$ of $\mathfrak{C}(\mathfrak{N})$ into $\mathfrak{C}(\mathfrak{M})$ is a contravariant functor of Φ.

Ex. 2. Let Φ be an analytic homeomorphism of $\mathfrak{M}$ onto $\mathfrak{N}$, $d\Phi$ the corresponding isomorphism of $\mathfrak{L}(\mathfrak{M})$ onto $\mathfrak{L}(\mathfrak{N})$, and $(d\Phi)^*$ the transpose of $d\Phi$, as defined in 1.5. Then $(d\Phi)^*$ coincides with the mapping $\{\Phi\}$ on $\mathfrak{L}^*(\mathfrak{N})$ defined by (10).

Ex. 3. Prove the invariance of the exterior multiplication by using the homomorphism property of $\{\Phi\}$. (A direct proof of this fact is implicit in the argument at the end of 4.2.)

4.4. Maurer-Cartan forms. Since the linear differential forms are a kind of dual of the infinitesimal transformations, we may again specialize by taking the manifold to be a Lie group $\mathfrak{G}$, and ask for the left-invariant differential forms on $\mathfrak{G}$.

In $\mathfrak{G}$, the left translation λ_a defines a mapping

$$d\lambda_a\colon \mathfrak{L}_x \to \mathfrak{L}_{ax}$$

between the spaces of tangent vectors. The transpose of this mapping is

$$(d\lambda_a)^*\colon \mathfrak{L}^*_{ax} \to \mathfrak{L}^*_x.$$

We define $\delta\lambda_a$ to be the mapping $(d\lambda_{a^{-1}})^*$, thus $\delta\lambda_a$ maps $\mathfrak{L}^*_x$ into $\mathfrak{L}^*_{ax}$. Explicitly, if $\omega \in \mathfrak{L}^*(\mathfrak{G})$, then $\omega^{\delta\lambda_a}$ is given by

$$\langle X, \omega^{\delta\lambda_a}\rangle_{ax} = \langle X^{d\lambda_{a^{-1}}}, \omega\rangle_x. \tag{13}$$

This equation may also be taken as a definition of $\delta\lambda_a$, since the scalar products with the $X \in \mathfrak{L}(\mathfrak{G})$ determine ω completely; this remains true even if we restrict X in (13) to be left-invariant. For the left-invariant X's (the elements of the Lie algebra of $\mathfrak{G}$) still span the space of tangent vectors $\mathfrak{L}_a$ at each point a of $\mathfrak{G}$. A further advantage of putting the definition in the form (13) is that it then applies to any linear differential form, analytic or not.—We note that the mapping $\delta\lambda_a$ could also have been defined as $\{\lambda_{a^{-1}}\}$, in the sense of 4.3 (cf. 4.3, Ex. 2), but we shall not make use of this fact.

A linear differential form ω on $\mathfrak{G}$ is called *left-invariant* or a *Maurer-Cartan form, MC-form* for short, if

$$\omega^{\delta\lambda_a} = \omega \quad \text{for all } a \in \mathfrak{G}.$$

Such a form is completely determined by its value at any one point. For if we know ω at $x \in \mathfrak{G}$, and $\omega = \omega^{\delta\lambda_a}$, then (13) determines ω at ax. As a convenient criterion for MC-forms we have

THEOREM 4.4.1. *A linear differential form ω on $\mathfrak{G}$ is an* MC-*form if and only if $\langle X, \omega\rangle$ is constant on $\mathfrak{G}$ for each X of the Lie algebra $\mathfrak{g}$ of $\mathfrak{G}$.*

Proof. If ω is an MC-form, and $X \in \mathfrak{g}$, then

$$\langle X, \omega \rangle_x = \langle X^{d\lambda_a}, \omega^{\delta\lambda_a} \rangle_{ax} = \langle X, \omega \rangle_{ax} \quad \text{for all } a, x \in \mathfrak{G},$$

hence $\langle X, \omega \rangle$ is constant on $\mathfrak{G}$. Conversely, let $\langle X, \omega \rangle$ be constant on $\mathfrak{G}$ for each $X \in \mathfrak{g}$. Then $\langle X, \omega \rangle_x = \langle X, \omega \rangle_{a^{-1}x} = \langle X, \omega^{\delta\lambda_a} \rangle_x$ for each $X \in \mathfrak{g}$, hence

$$\langle X, \omega - \omega^{\delta\lambda_a} \rangle_x = 0.$$

Now the X in $\mathfrak{g}$, taken at the point x of $\mathfrak{G}$, span the whole space $\mathfrak{L}_x$, whence $\omega - \omega^{\delta\lambda_a} = 0$ at $x \in \mathfrak{G}$. This holds for all $x \in \mathfrak{G}$, therefore $\omega = \omega^{\delta\lambda_a}$, and since this is true for all $a \in \mathfrak{G}$, ω is an MC-form, as we wished to prove.

THEOREM 4.4.2. *Any* MC-*form on a Lie group* $\mathfrak{G}$ *is analytic, and the set of all such forms on* $\mathfrak{G}$ *is a vector space of dimension equal to the dimension of the group.*

We begin by proving the second part. Let $\mathfrak{g}$ be the Lie algebra of $\mathfrak{G}$, and $X_1, \ldots, X_n$ a basis of $\mathfrak{g}$; we define n linear differential forms ω^j by the equations

$$\langle X_i, \omega^j \rangle = \delta_i^j.$$

Since the X_i are linearly independent at each point of $\mathfrak{G}$, this defines the ω^j uniquely, by Theorem 1.5.1. They are MC-forms by the criterion we have just proved (Theorem 4.4.1), and clearly any linear combination with constant coefficients is again an MC-form. Conversely if ω is any MC-form on $\mathfrak{G}$, then

$$\omega = \langle X_i, \omega \rangle \, \omega^i,$$

and, again by Theorem 4.4.1, this expresses ω as a linear combination, with constant coefficients, of the ω^i.

To prove that any MC-form is analytic it is enough, by linearity, to prove this for the ω^i. Let us take a chart at e and write

$$X_i = \psi_i^j(x) \frac{\partial}{\partial x^j};$$

since both X_i and $\partial/\partial x^i$ form a basis for the space $\mathfrak{L}_e(\mathfrak{G})$ of tangent vectors at e, the matrix (ψ_i^j) is non-singular at e, and therefore by continuity in some neighbourhood of e. In this neighbourhood it must then have an inverse, $(\bar{\psi}_i^j)$ say, which is analytic by Theorem A.2 of the Appendix. Now consider the forms $\varpi^i = \bar{\psi}^i_j(x)\,dx^j$. These forms are analytic at e and satisfy

$$\begin{aligned}\langle X_i, \varpi^j\rangle &= \left\langle \psi^k_i \frac{\partial}{\partial x^k}, \bar{\psi}^j_l\,dx^l \right\rangle \\ &= \psi^k_i \bar{\psi}^j_l \left\langle \frac{\partial}{\partial x^k}, dx^l \right\rangle \\ &= \psi^k_i \bar{\psi}^j_l\, \delta^l_k = \psi^k_i \bar{\psi}^j_k = \delta^j_i.\end{aligned}$$

By uniqueness (Theorem 1.5.1), $\omega^i = \varpi^i$; hence the ω^i are analytic at e, and by translation everywhere on $\mathfrak{G}$. This completes the proof.

These two theorems justify us in regarding the space of MC-forms as the dual of the space $\Lambda(\mathfrak{G})$ of infinitesimal right translations, and we therefore denote the space of MC-forms on $\mathfrak{G}$ by $\Lambda^*(\mathfrak{G})$.

In the course of the proof we have obtained an explicit expression for the forms ω^i; we state this as a

Corollary. *If $X_i = \psi^j_i . \partial/\partial x^j$ is a basis of the Lie algebra $\Lambda(\mathfrak{G})$ of $\mathfrak{G}$ in terms of some chart at e, then the forms $\omega^i = \bar{\psi}^i_j dx^j$, where $(\bar{\psi}^i_j)$ is the inverse of the matrix (ψ^j_i), constitute a basis of $\Lambda^*(\mathfrak{G})$.*

We can extend the concept of the inner product $\langle X, \omega\rangle$ to forms of higher degree as follows: If $X_1, \ldots, X_k$ are any infinitesimal transformations, we write

$$\langle X_1, \ldots, X_k;\, dx^1 \wedge \ldots \wedge dx^k\rangle = \det(\langle X_i, dx^j\rangle),$$

and extend this definition to any homogeneous forms of degree k by linearity. The definition amounts just to regarding a homogeneous differential form as an alternating multilinear form on the space of infinitesimal transformations.

The linear mapping $\delta\lambda_a$ of $\mathfrak{L}^*(\mathfrak{G})$ into itself can be extended, as in 4.3, to a homomorphism of the exterior algebra $\mathfrak{C}(\mathfrak{G})$ of all analytic differential forms on $\mathfrak{G}$ into itself, i.e. an endomorphism of $\mathfrak{C}(\mathfrak{G})$. Any form $\omega \in \mathfrak{C}(\mathfrak{G})$ which satisfies $\omega^{\delta\lambda_a} = \omega$ for all $a \in \mathfrak{G}$ is then called an MC-form, and it can be shown that we lose nothing here by restricting our attention to analytic forms. As is easily verified, a homogeneous k-linear form ω is an MC-form if and only if $\langle X_1, \ldots, X_k;\, \omega\rangle$ is constant on $\mathfrak{G}$ for any choice of $X_1, \ldots, X_k$ in $\mathfrak{g}$. It follows from this fact that the set of MC-forms on $\mathfrak{G}$ is just the exterior algebra on the vector space $\Lambda^*(\mathfrak{G})$. We shall denote this algebra by $\mathfrak{E}(\mathfrak{G})$.† Thus $\mathfrak{E}(\mathfrak{G})$ consists of all MC-forms on $\mathfrak{G}$, and is an algebra of dimension 2^n, where n is the dimension of $\mathfrak{G}$.

If we observe that the mapping $\delta\lambda_a$ is just the mapping $\{\lambda_{a^{-1}}\}$ as defined in 4.3, we may apply Lemma 4.3.2 to show that the algebra $\mathfrak{E}(\mathfrak{G})$ is mapped into itself by exterior differentiation. But we shall give another proof of this fact in Theorem 4.4.3, by means of a precise formula.

THEOREM 4.4.3. *If* $\mathfrak{G}$ *is any Lie group and* $X, Y \in \Lambda(\mathfrak{G})$, $\omega \in \Lambda^*(\mathfrak{G})$, *then*

$$\langle X, Y; d\omega\rangle + \langle [X, Y], \omega\rangle = 0. \tag{14}$$

Proof. It follows from the hypothesis that $\langle X, \omega\rangle$ and $\langle Y, \omega\rangle$ are constant on $\mathfrak{G}$, and this is in fact all we shall use. Suppose that $\omega = \alpha_i dx^i$ in terms of a chart. Then

$$\langle X, \omega\rangle = \langle X, \alpha_i dx^i\rangle = \alpha_i\langle X, dx^i\rangle = \alpha_i Xx^i.$$

Since $\langle X, \omega\rangle$ is constant, we have $Y\langle X, \omega\rangle = 0$, i.e.

$$0 = Y(\alpha_i . Xx^i) = Y\alpha_i . Xx^i + \alpha_i . YXx^i.$$

Similarly $$0 = X\alpha_i . Yx^i + \alpha_i . XYx^i;$$

by subtraction we find

$$X\alpha_i . Yx^i - Y\alpha_i . Xx^i + \alpha_i[X, Y]x^i = 0,$$

i.e.

$$\langle X, d\alpha_i\rangle\langle Y, dx^i\rangle - \langle Y, d\alpha_i\rangle\langle X, dx^i\rangle + \langle [X, Y], \alpha_i dx^i\rangle = 0.$$

The first two terms are by definition $\langle X, Y; d\alpha_i \wedge dx^i\rangle$, hence

$$\langle X, Y; d\omega\rangle + \langle [X, Y], \omega\rangle = 0,$$

as was to be proved.

We note that the proof applies to any analytic infinitesimal transformations X, Y and analytic linear differential form ω on an analytic manifold $\mathfrak{M}$ such that $\langle X, \omega\rangle$ and $\langle Y, \omega\rangle$ are constant on $\mathfrak{M}$.

† $\mathfrak{E}(\mathfrak{G})$ is also known as the *Grassmann algebra* on the Lie algebra $\mathfrak{g}$.

Ex. Let $\mathfrak{G}$ be a Lie group. If $X_1, \ldots, X_{r+1} \in \Lambda(\mathfrak{G})$ and ω is an r-linear MC-form on $\mathfrak{G}$, then†

$$\langle X_1, \ldots, X_{r+1}; d\omega\rangle = \sum_{i<j} (-1)^{i+j} \langle [X_i, X_j], X_1, \ldots, \hat{X}_i, \ldots, \hat{X}_j, \ldots, X_{r+1}; \omega\rangle, \quad (15)$$

where the sign ^ over a letter indicates that this letter is to be omitted.

4.5. The Maurer-Cartan relations. If ω is any linear MC-form, then by Theorem 4.4.3, $d\omega$ is again an MC-form and can therefore be expressed in terms of a basis of $\mathfrak{E}(\mathfrak{G})$, the algebra of MC-forms. We shall now derive an explicit formula for this expression.

Let $X_1, \ldots, X_n$ be a basis of the Lie algebra $\mathfrak{g}$ of $\mathfrak{G}$. Then

$$[X_i, X_j] = c_{ij}^k X_k, \quad (16)$$

where the c_{ij}^k are the structure constants of the group. Now let $\omega^1, \ldots, \omega^n$ be the basis dual to $X_1, \ldots, X_n$, defined by

$$\langle X_i, \omega^j\rangle = \delta_i^j.$$

The ω^i generate $\mathfrak{E}(\mathfrak{G})$, and by Theorem 4.4.3 $d\omega^k$ is an MC-form. Therefore it belongs to $\mathfrak{E}(\mathfrak{G})$ and so we have

$$d\omega^k = \gamma_{ij}^k \omega^i \wedge \omega^j, \quad (17)$$

where the γ_{ij}^k are certain constants. If we apply (16) and (17) to the formula

$$\langle X_r, X_s; d\omega^t\rangle + \langle [X_r, X_s], \omega^t\rangle = 0$$

obtained from Theorem 4.4.3, we find

$$\langle X_r, X_s; \gamma_{ij}^t \omega^i \wedge \omega^j\rangle + \langle c_{rs}^k X_k, \omega^t\rangle = 0.$$

Hence
$$\gamma_{ij}^t(\delta_r^i \delta_s^j - \delta_s^i \delta_r^j) + c_{rs}^k \delta_k^t = 0,$$

i.e.
$$\gamma_{rs}^t - \gamma_{sr}^t + c_{rs}^t = 0.$$

Thus $c_{rs}^t = -(\gamma_{rs}^t - \gamma_{sr}^t)$. Now the coefficient of $\omega^r \wedge \omega^s$ in (17) is just $\gamma_{rs}^t - \gamma_{sr}^t$. We have therefore

$$d\omega^k = -\sum_{\substack{r,s=1 \\ r<s}}^{n} c_{rs}^k \omega^r \wedge \omega^s.$$

† Equation (15) is sometimes known as the coboundary formula. It can be used to define $d\omega$ for any alternating r-linear function on an abstract Lie algebra.

We can write this as a double sum without restriction, by using the skew-symmetry of $\omega^r \wedge \omega^s$:

$$d\omega^k = -\tfrac{1}{2} c^k_{ij} \omega^i \wedge \omega^j. \tag{18}$$

These formulae (18) are known as the *Maurer-Cartan relations*. They express the duality between the Lie algebra and the algebra of MC-forms. In particular, it can be shown that the Jacobi identity corresponds to the identity $d^2\omega = 0$ in the algebra of MC-forms. This shows that the Jacobi identity corresponds in some sense to an integrability condition (cf. Chapter V).

In the derivation of (18) we used only the fact that the X_i form a set of infinitesimal transformations satisfying (16); moreover, the differential forms ω^j are uniquely determined by the equations $\langle X_i, \omega^j \rangle = \delta^j_i$, provided that at each point $p \in \mathfrak{M}$ the tangent vectors defined by the X_i form a basis of $\mathfrak{L}_p(\mathfrak{M})$ (cf. Theorem 1.5.1). Thus we can sum up the result as

THEOREM 4.5.1. *Let $X_1, \ldots, X_n$ be n analytic infinitesimal transformations on an analytic manifold $\mathfrak{M}$ satisfying* (16), *and suppose that their tangent vectors at each point p of $\mathfrak{M}$ form a basis of $\mathfrak{L}_p(\mathfrak{M})$. Then the linear differential forms ω^j defined by $\langle X_i, \omega^j \rangle = \delta^j_i$ are uniquely determined and satisfy*

$$d\omega^k = -\tfrac{1}{2} c^k_{ij} \omega^i \wedge \omega^j.$$

The Maurer-Cartan relations may be brought to a particularly simple form if we introduce formal products of differential forms by infinitesimal transformations.

If $X_1, \ldots, X_n$ is again a basis of the Lie algebra $\mathfrak{g}$ of $\mathfrak{G}$, we may consider expressions of the form

$$\Theta = \theta^i X_i, \tag{19}$$

where the θ^i are any MC-forms, not necessarily linear. Two such generalized forms $\theta^i X_i$ and $\chi^i X_i$ are said to be *equal* if and only if $\theta^i = \chi^i$ $(i = 1, \ldots, n)$, and we define addition and multiplication by the formulae†

$$\alpha(\theta^i X_i) + \beta(\chi^i X_i) = (\alpha\theta^i + \beta\chi^i) X_i \quad (\alpha, \beta \in R),$$
$$\theta^i X_i \wedge \chi^j X_j = \theta^i \wedge \chi^j [X_i, X_j].$$

† These formulae may be expressed by saying that we have formed the tensor product (over R) of the Lie algebra $\mathfrak{g}$ and the algebra $\mathfrak{E}(\mathfrak{G})$.

We shall say that $\Theta = \theta^i X_i$ is linear (bilinear, etc.) if each θ^i is a linear (bilinear, etc.) differential form, and we define the operation d by the equation

$$d(\theta^i X_i) = d\theta^i X_i.$$

Now we consider the element

$$\Omega = \omega^i X_i,$$

where ω^i is the basis of MC-forms dual to the X_i. We have

$$\begin{aligned} d\Omega = d\omega^k X_k &= -\tfrac{1}{2} c^k_{ij} \omega^i \wedge \omega^j X_k \\ &= -\tfrac{1}{2} \omega^i \wedge \omega^j [X_i, X_j] \\ &= -\tfrac{1}{2} \Omega \wedge \Omega. \end{aligned}$$

Thus the Maurer-Cartan relations take on the form

$$d\Omega = -\tfrac{1}{2} \Omega \wedge \Omega. \tag{20}$$

Ex. 1. If Θ, X are two generalized MC-forms, which are homogeneous of degrees r, s respectively, then

$$\mathrm{X} \wedge \Theta = (-1)^{rs+1} \Theta \wedge \mathrm{X}.$$

Ex. 2. If f is any analytic function on a Lie group $\mathfrak{G}$, then

$$df = X_i f . \omega^i,$$

where X_i is a basis of the Lie algebra of $\mathfrak{G}$ and ω^i is the dual basis of MC-forms. This generalizes the equation $df = \dfrac{\partial f}{\partial x^i} dx^i$.

Ex. 3. For the affine group in one variable, determine a basis of $\Lambda^*(\mathfrak{G})$ dual to the basis of $\Lambda(\mathfrak{G})$ found in 3.3, Ex. 1, and verify (18).

CHAPTER V

LIE'S FUNDAMENTAL THEOREMS

5.1. Statement of the theorems. In Chapter III we have seen that to every Lie group there corresponds a Lie algebra; we shall now show that, conversely, to every Lie algebra there corresponds a local Lie group. The remaining question whether we can pass to a global Lie group will be considered later, though we shall not give a complete answer to this question.

We recall the following notations: The composition functions in terms of a chart at e are given by $(xy)^i = \phi^i(x, y)$; we write again

$$\psi^i_j(x) = \left[\frac{\partial \phi^i(x, y)}{\partial y^j}\right]_{y=e} \tag{1}$$

for the transformation functions, and denote the inverse of the matrix $(\psi^i_j(x))$ by $(\check{\psi}^i_j(x))$. Further we define

$$\chi^i_j(xy, y) = \frac{\partial \phi^i(x, y)}{\partial y^j}. \tag{2}$$

If we put $xy = z$, then (2) just states that $\chi^i_j(z, y) = \partial z^i / \partial y^j$, where zy^{-1} is kept constant during the differentiation.

We note the following formulae:

$$\chi^i_j(x, e) = \psi^i_j(x), \tag{3}$$

$$\chi^i_j(x, x) = \delta^i_j. \tag{4}$$

Equation (3) is simply the definition of ψ^i_j, and to obtain (4) we put $x = e$ in (2) and replace y by x.

Lemma 5.1.1. *For any x, y and z near e,*

$$\chi^i_k(x, z) = \chi^i_j(x, y)\, \chi^j_k(y, z).$$

Proof. Write $x = uy$, $y = vz$; then $x = uvz$, or in terms of coordinates, $x^i = \phi^i(u, \phi(v, z))$. Applying the rule for differentiating a function of a function, we find

$$\begin{aligned} \chi^i_k(x, z) &= \frac{\partial \phi^i(u, y)}{\partial y^j} \frac{\partial \phi^j(v, z)}{\partial z^k} \\ &= \chi^i_j(x, y)\, \chi^j_k(y, z), \end{aligned}$$

which is the required formula.

COROLLARY 1. $\chi^i_j(e,x) = \breve{\psi}^i_j(x)$.

For we have $\chi^i_j(e,x)\,\chi^j_k(x,e) = \chi^i_k(e,e) = \delta^i_k$ by the lemma and (4), hence $\chi^i_j(e,x)\,\psi^j_k(x) = \delta^i_k$, by (3); this equation determines the inverse matrix uniquely and so the corollary follows.

COROLLARY 2. $\chi^i_k(x,y) = \psi^i_j(x)\,\breve{\psi}^j_k(y)$.

For, by the lemma

$$\begin{aligned}\chi^i_k(x,y) &= \chi^i_j(x,e)\,\chi^j_k(e,y)\\ &= \psi^i_j(x)\,\breve{\psi}^j_k(y),\end{aligned}$$

where we have used (3) and Corollary 1.

We note that, conversely, Lemma 5.1.1 can be deduced from Corollary 2. Corollary 2 is usually stated as

Lie's first theorem. *If $z^i = \phi^i(x,y)$ are the composition functions of a Lie group, then*

$$\frac{\partial z^i}{\partial y^k} = \psi^i_j(z)\,\breve{\psi}^j_k(y),$$

where $(\psi^i_j(z))$ is a matrix of analytic functions which reduces to (δ^i_j) for some value of the arguments z^k, and which has the inverse $(\breve{\psi}^i_j(z))$ in some neighbourhood of this value.

The content of Lie's first theorem can also be expressed in terms of infinitesimal right translations or in terms of MC-forms.

Let f be any analytic function in $\mathfrak{G}$, defined near e, and consider f as a function of $z = xy$. Then

$$\frac{\partial f}{\partial y^k} = \frac{\partial f}{\partial z^i}\frac{\partial \phi^i(x,y)}{\partial y^k} = \frac{\partial f}{\partial z^i}\chi^i_k(z,y).$$

By Corollary 2 of Lemma 5.1.1 this may be written

$$\frac{\partial f}{\partial y^k} = \frac{\partial f}{\partial z^i}\psi^i_j(z)\,\breve{\psi}^j_k(y). \tag{5}$$

Multiplying by $\psi^k_l(y)$, we obtain

$$\psi^k_l(y)\frac{\partial f}{\partial y^k} = \psi^i_l(z)\frac{\partial f}{\partial z^i}. \tag{6}$$

These equations are the analytical expression of the fact that the infinitesimal right translations $\psi^k_l(y)\dfrac{\partial}{\partial y^k}$ are invariant under the left translation $\lambda_x\colon y \to xy$.

Next we have, by (5),

$$df = \frac{\partial f}{\partial y^k}\, dy^k = \frac{\partial f}{\partial z^i}\, \psi^i_j(z)\, \bar{\psi}^j_k(y)\, dy^k.$$

If we choose $f = z^h$, then

$$dz^h = \psi^h_j(z)\, \bar{\psi}^j_k(y)\, dy^k,$$

and multiplication by $\bar{\psi}^i_h(z)$ yields

$$\bar{\psi}^i_h(z)\, dz^h = \bar{\psi}^i_k(y)\, dy^k. \tag{7}$$

This expresses the left-invariance of the MC-forms. The equations (7) are sometimes known as the *Maurer-Cartan equations*. We note that each of the sets (6), (7) of equations is equivalent to Lie's first theorem.

For reference we state Lie's other theorems.

Lie's second theorem. *If $X_1, \ldots, X_n$ are the infinitesimal transformations of a Lie group, then the coefficients c^k_{ij} in the equations*

$$[X_i, X_j] = c^k_{ij} X_k$$

are constants.

This was proved in 3.2; it is the essential content of Theorem 3.2.3.

Lie's third theorem. *The structure constants satisfy*

$$c^k_{ij} + c^k_{ji} = 0,$$

$$c^r_{ij} c^s_{rk} + c^r_{jk} c^s_{ri} + c^r_{ki} c^s_{rj} = 0.$$

These equations are equivalent to the defining relations of a Lie algebra: $[X, X] = 0$, $[[X, Y], Z] + [[Y, Z], X] + [[Z, X], Y] = 0$, which were proved in Theorem 3.1.2.

5.2. The converses of Lie's first and second theorems.

For the converse of Lie's theorems we have to consider the integration of a system of total differential equations; we recall briefly how such systems arise. Since all the results in this chapter are purely local, we may without loss of generality operate in Euclidean space.

Suppose that we have a curve in N-dimensional space. By choosing a suitable coordinate system

$$x^1, \ldots, x^n, t \quad (n = N - 1),$$

we may write the curve as

$$x^i = \phi^i(t) \quad (i = 1, \ldots, n). \tag{8}$$

Under suitable differentiability assumptions (8) is the unique solution of a system of differential equations†

$$\frac{dx^i}{dt} = \gamma^i(t, x) \quad (x^i = a^i \text{ when } t = 0). \tag{9}$$

Conversely, for a given set of analytic (or even continuously differentiable) functions $\gamma^i(t, x)$, the system (9) has a unique solution, which is of the form (8).

In the theory of total differential equations we have the generalization of this process to the case of surfaces. Consider an r-dimensional surface Σ in N-dimensional space and write $n = N - r$. Then in a suitable coordinate system $x^1, \ldots, x^n, t^1, \ldots, t^r$ we can write Σ in the form

$$x^i = \phi^i(t) \quad (\equiv \phi^i(t^1, \ldots, t^r)). \tag{10}$$

We want to set up a system of differential equations for the ϕ^i. A system corresponding to the 1-dimensional case would be

$$\frac{\partial x^i}{\partial t^\alpha} = \gamma^i_\alpha(t, x), \tag{11}$$

with the initial conditions $x^i = a^i$ when $t^\alpha = 0$.‡ Such a system always exists (and again can be chosen in many ways) if the ϕ's are suitably differentiable, and it has at most one solution; but a given system may have no solution at all. For example, when $n = 1$, $r = 2$, the equations $\partial x/\partial s = t$, $\partial x/\partial t = 0$ have no solution $x = \phi(s, t)$, for such a solution would have to satisfy

$$\frac{\partial^2 \phi}{\partial s \, \partial t} = 0, \quad \frac{\partial^2 \phi}{\partial t \, \partial s} = 1,$$

which is impossible if ϕ is to have continuous second derivatives.

Generally the equation $\dfrac{\partial^2 \phi}{\partial s \, \partial t} = \dfrac{\partial^2 \phi}{\partial t \, \partial s}$ can be expressed as a condition on the functions γ^i_α in (11) and this gives rise to certain integrability conditions on (11); whenever these are satisfied, (11) has a solution.§

† Of course there are many such systems determining a given curve (8), in other words, the correspondence (8)↔(9) is one-many.

‡ We shall denote suffixes in the range $1 \ldots r$ by Greek letters. Accordingly a repeated Greek suffix indicates a summation from 1 to r.

§ These integrability conditions are in fact necessary and sufficient, but we shall only require the sufficiency and we therefore omit the proof of the necessity.

To state the integrability conditions succinctly we write (11) as $dx^i = \gamma^i_\alpha(t,x)\,dt^\alpha$, or more briefly,

$$dx^i = \Gamma^i(t,x,dt), \tag{12}$$

where $\Gamma^i = \gamma^i_\alpha(t,x)\,dt^\alpha$. We may interpret (12) as follows: at any point of Σ, (12) gives the approximate change in x^i, when we change t^α by a small amount dt^α, keeping in the surface Σ.

THEOREM 5.2.1. *Let $\gamma^i_\alpha(t,x)$ be nr functions of $r+n$ variables t^α, x^i $(\alpha = 1,\ldots,r; i = 1,\ldots,n)$, analytic at $t^\alpha = 0$, $x^i = 0$, and write $\Gamma^i(t,x,dt) = \gamma^i_\alpha(t,x)\,dt^\alpha$. Then the system*

$$dx^i = \Gamma^i(t,x,dt) \quad (x^i = a^i \text{ when } t^\alpha = 0) \tag{13}$$

has a unique solution $x^i = \phi^i(t,a)$ analytic at $t^\alpha = 0$, $a^i = 0$, provided that

$$d\Gamma^i(t,x,dt) = 0 \quad \text{near} \quad t^\alpha = 0,\ x^i = 0. \tag{14}$$

Here (14) is to be understood in the sense that in $d\Gamma^i$, each dx^i is expressed in terms of the dt's by means of (13).

For the proof we refer to the appendix (Theorem A 1′).

Now we are in a position to prove the converses of Lie's first two theorems.

THEOREM 5.2.2 (converse of Lie's first theorem). *Let n functions of 2n variables $x^1,\ldots,x^n,y^1,\ldots,y^n$ be given,*

$$z^i = \phi^i(x,y), \tag{15}$$

where the ϕ's are analytic at $x^i = 0$, $y^j = 0$. Suppose that they satisfy the equations

$$\frac{\partial z^i}{\partial y^k} = \psi^i_j(z)\,\bar\psi^j_k(y) \quad (z^i = x^i \text{ when } y^j = 0), \tag{16}$$

where the functions $\psi^i_j(z)$ are analytic at $z^k = 0$, and satisfy

$$\psi^i_j(0) = \delta^i_j.$$

Further suppose that $(\bar\psi^i_j)$ is the inverse of the matrix (ψ^i_j).

Then the functions ϕ^i are the composition functions of a local Lie group with infinitesimal transformations $X_i = \psi^j_i(x)\dfrac{\partial}{\partial x^j}$.

Proof. We denote the point of R^n with the coordinates x^i by x, and write xy and e for the points with the coordinates $\phi^i(x,y)$

nd 0 respectively. The initial conditions in (16) can then be vritten

$$xe = x. \tag{17}$$

The equations (16) also show that $[\partial z^i/\partial y^k]_{y=e} = \psi^i_j(x)\,\bar{\psi}^j_k(e)$, and ience $[\partial z^i/\partial y^k]_{x=e,\,y=e} = \delta^i_k$; this proves that

$$\det\left[\frac{\partial \phi^i}{\partial y^k}\right]_{x=e,\,y=e} = 1 \neq 0.$$

To prove that the ϕ^i define a local Lie group it only remains for us to verify the associative law: $(xy)z = x(yz)$.

Let us write $v = yz$, $w = xv$, then by (16),

$$dv^i = \psi^i_j(v)\,\bar{\psi}^j_k(z)\,dz^k \quad \text{and} \quad dw^i = \psi^i_j(w)\,\bar{\psi}^j_k(v)\,dv^k,$$

hence

$$\begin{aligned} dw^i &= \psi^i_j(w)\,\bar{\psi}^j_k(v)\,\psi^k_l(v)\,\bar{\psi}^l_p(z)\,dz^p \\ &= \psi^i_j(w)\,\bar{\psi}^j_p(z)\,dz^p, \end{aligned}$$

by the definition of $\bar{\psi}^i_j$. When $z = e$, we have $w = xv = x(ye) = xy$ (using (17)); thus $w = x(yz)$ satisfies the same equation as $(xy)z$ and the same initial conditions. Therefore, by the uniqueness of the solution, $(xy)z = x(yz)$.

To complete the proof we calculate the transformation functions of the local Lie group defined by (15):

$$\left[\frac{\partial \phi^i(x,y)}{\partial y^k}\right]_{y=e} = [\psi^i_j(xy)\,\bar{\psi}^j_k(y)]_{y=e} = \psi^i_k(x).$$

Hence the $\psi^i_k(x)\dfrac{\partial}{\partial x^i}$ are the infinitesimal transformations of the local Lie group, and the proof is complete.

In Theorem 5.2.2 we were given certain functions $\phi^i(x, y)$ and proved that they defined a group because they satisfied differential equations of a certain kind. At the next stage we take a set of infinitesimal transformations forming a basis of a Lie algebra and from it construct a local Lie group.

THEOREM 5.2.3 (converse of Lie's second theorem). *Let Λ be a Lie algebra of analytic infinitesimal transformations defined on an open non-empty subset V of R^n. Suppose further*

(i) *the algebra Λ is of dimension n over the field R,*

(ii) *at some point p of V the space Λ_p of tangent vectors at p defined by Λ has dimension n.*

Then there is a local Lie group, unique to within local analytic isomorphism, with Λ as its Lie algebra.

Proof. We take p as origin of coordinates in R^n and choose a basis X_i of Λ which reduces to $\partial/\partial x^i$ at p. In terms of these coordinates we have

$$X_i = \psi_i^j(x)\frac{\partial}{\partial x^j},$$

say, where the ψ_i^j are analytic in V, and $\psi_i^j(0) = \delta_i^j$. The matrix (ψ_i^j) is non-singular at 0, therefore also in some neighbourhood W of 0, and so it has an inverse $(\bar{\psi}_i^j)$ in W, where the $\bar{\psi}$'s are analytic in W (Theorem A2). If we define n linear differential forms ω^i on W by the equations $\omega^i = \bar{\psi}_j^i(x)\,dx^j$, then $\langle X_i, \omega^j\rangle = \delta_i^j$, and so, by Theorem 4.5.1,

$$d\omega^k = -\tfrac{1}{2}c_{ij}^k\,\omega^i \wedge \omega^j, \tag{18}$$

where the c_{ij}^k are the multiplication constants of the algebra Λ. To complete the proof we need only show that the equations

$$dz^i = \psi_j^i(z)\,\bar{\psi}_k^j(y)\,dy^k \tag{19}$$

satisfy the integrability condition when y^i, z^j are near 0. For if $z^i = \phi^i(x, y)$ is the solution of (19) which reduces to x^i for $y^j = 0$, then by Theorem 5.2.2 these equations define a local Lie group with the infinitesimal transformations X_i; its Lie algebra is therefore Λ. The uniqueness follows from the fact that in a suitable neighbourhood of 0 the equations (19) have a unique solution.

Let us write

$$\Phi^i \equiv \Phi^i(y, z, dy, dz) = dz^i - \psi_j^i(z)\,\bar{\psi}_k^j(y)\,dy^k.$$

Then the integrability condition for (19) may be stated thus: If we calculate $d\Phi^i$ and use the equations $\Phi^i = 0$ to express dz^k in terms of the dy's, we must obtain an expression which vanishes identically.

This condition may be transformed as follows: Let $\alpha_j^i(y, z)$ be any set of n^2 analytic functions of the y's and z's whose determinant does not vanish near 0. Then

$$d(\alpha_j^i\Phi^j) = d\alpha_j^i \wedge \Phi^j - \alpha_j^i\,d\Phi^j.$$

Hence, under the hypothesis that $\Phi^i = 0$,

$$d(\alpha_j^i\Phi^j) = 0 \ (i = 1, \ldots, n) \text{ if and only if } d\Phi^i = 0 \ (i = 1, \ldots, n).$$

We choose $\alpha^i_j = \psi^i_j(z)$. Then

$$\psi^i_j(z)\,\Phi^j = \psi^i_j(z)\,dz^j - \psi^i_k(y)\,dy^k$$
$$= \omega^i(z) - \omega^i(y) \quad \text{say,}$$

where ω^i satisfies (18). Hence the equations (19) are equivalent to

$$\omega^k(z) - \omega^k(y) = 0. \tag{20}$$

Applying d we obtain

$$d\omega^k(z) - d\omega^k(y) = -\tfrac{1}{2}c^k_{ij}\{\omega^i(z) \wedge \omega^j(z) - \omega^i(y) \wedge \omega^j(y)\}$$
$$= 0$$

by (20). Thus the integrability is assured and the theorem follows.

Ex. 1. Prove that the condition (14) of Theorem 5.2.1 is necessary for (13) to have a solution $x^i = \phi^i(t, a)$, analytic at $t^\alpha = 0$, $a^i = 0$.

Ex. 2. Show that the three hypotheses of Theorem 5.2.3—(1) Λ is a Lie algebra, (2) $\dim \Lambda = n$, (3) $\dim \Lambda_p = n$ at some point p—are independent.

5.3. The integration of bilinear differential forms.

In order to construct the local Lie group from its infinitesimal transformations we had to integrate a system of differential equations for the functions $\phi^i(x, y)$, which were essentially a system of linear differential forms. In the converse of the third theorem we are given an abstract Lie algebra and we have to find its infinitesimal transformations or what is equivalent, its MC-forms. Thus we are given the multiplication constants c^k_{ij} of the Lie algebra and we have to solve the equations

$$d\omega^k = -\tfrac{1}{2}c^k_{ij}\,\omega^i \wedge \omega^j \quad (\omega^k = dx^k \text{ when } x^i = 0).$$

This is a system of bilinear differential forms and we shall prove that it has a solution whenever the integrability conditions—which express the fact that $d^2\omega^k = 0$—are satisfied.

THEOREM 5.3.1. *Let $\Gamma^i(\omega, dt)$ be n bilinear expressions in the linear differential forms $\omega^1, \ldots, \omega^n, dt^1, \ldots, dt^r$, with constant coefficients.*

Then the equations

$$\left.\begin{aligned} d\omega^k &= \Gamma^k(\omega, dt), \\ \omega^k &= c^k_\alpha dt^\alpha \quad \textit{when } t^\beta = 0, \end{aligned}\right\} \tag{21}$$

have a solution which is analytic at $t^\beta = 0$, $c^k_\beta = 0$, *provided that*

$$d\Gamma^k(\omega, dt) = 0; \tag{22}$$

here (22) *is to hold identically in the* ω^i *and* dt^α, *the* $d\omega^k$ *being replaced by their values from* (21).

We shall prove the theorem by induction on r, the number of variables t^α. When $r = 1$, each form ω^i is a multiple of dt^1, and so all the bilinear forms Γ^k reduce to 0. Therefore $\omega^k = c^k_1 dt^1$ is a solution in this case.

Now let $r > 1$ and consider the system obtained from (21) by keeping t^1 constant. This is a system in $r - 1$ variables, and by the induction hypothesis it has a solution

$$\omega^i = \sum_{\sigma=2}^{r} \phi^i_\sigma(t^2, \ldots, t^r)\, dt^\sigma, \tag{23}$$

where the ϕ^i_σ are analytic at $t^\rho = 0$, $c^k_\rho = 0$ $(k = 1, \ldots, n; \rho = 2, \ldots, r)$. If we allow t^1 to vary again and define $\phi^i_1 = 0$, we may write (23) as $\omega^i = \phi^i$, where $\phi^i = \phi^i_\alpha dt^\alpha$. In order to solve (21), we put

$$\omega^i = u^i dt^1 + \phi^i,$$

and try to determine u^i as a function of $t^2, \ldots, t^r$ such that ω^i satisfies (21). We show that this can be done by finding a system of total differential equations for the u^i which are integrable.

Both $d\phi^i$ and $\Gamma^i(u\,dt^1 + \phi, dt)$ are bilinear differential forms with coefficients which depend on $u^1, \ldots, u^n$ and $t^2, \ldots, t^r$ only. Thus

$$\Gamma^i(u\,dt^1 + \phi, dt) - d\phi^i = a^i_{\alpha\beta}(u, t)\, dt^\alpha \wedge dt^\beta. \tag{24}$$

When $dt^1 = 0$, the left-hand side vanishes, by construction. Hence

$$\sum_{\rho,\sigma=2}^{r} a^i_{\rho\sigma}(u, t)\, dt^\rho \wedge dt^\sigma = 0,$$

and (24) reduces to

$$\Gamma^i(u\,dt^1 + \phi, dt) - d\phi^i = A^i(u, t, dt) \wedge dt^1, \tag{25}$$

where A^i is an analytic linear differential form in $dt^2, \ldots, dt^r$.

Now if $u^i dt^1 + \phi^i$ is to satisfy (21), then

$$du^i \wedge dt^1 + d\phi^i = \Gamma^i(u\,dt^1 + \phi, dt),$$

and by a comparison with (25) we obtain

$$du^i \wedge dt^1 = A^i(u, t, dt) \wedge dt^1. \tag{26}$$

The A^i do not depend on t^1 or dt^1 and if the u^i are also to be independent of t^1, they must satisfy, by (26),

$$\left.\begin{aligned} du^i &= A^i(u, t, dt), \\ u^i &= c_1^i, \quad \text{when } t^\sigma = 0\ (\sigma = 2, \ldots, r). \end{aligned}\right\} \tag{27}$$

To see that these equations can be integrated we apply exterior differentiation to (25). The left-hand side vanishes, for $d(d\phi^i) = 0$ by Theorem 4.3.1 (ii) and $d\Gamma^i = 0$ by (22), which is part of the hypothesis. Thus

$$dA^i(u, t, dt) \wedge dt^1 = 0,$$

and since A^i and u^i are independent of t^1, we find that

$$dA^i(u, t, dt) = 0,$$

which is the integrability condition for (27).

If u^i is a solution of (27) and $\omega^i = u^i dt^1 + \phi^i$, then

$$\begin{aligned} d\omega^i &= du^i \wedge dt^1 + d\phi^i \\ &= A^i \wedge dt^1 + d\phi^i \qquad \text{by (26)}, \\ &= \Gamma^i(u\,dt^1 + \phi, dt) \quad \text{by (25)}. \end{aligned}$$

Clearly ω^i also satisfies the given initial condition, and it is therefore the required solution. This completes the proof.

We cannot expect to find a unique solution, for if, for example, in $d\omega^i = \Gamma^i(\omega, t, dt)$ the right-hand side does not depend on ω^1, we may, in any given solution, add to ω^1 any differential form ϕ which satisfies $d\phi = 0$ and which vanishes at $t^\alpha = 0$. However, when the functions Γ^i are independent of the t's and dt's, the solution is unique in the sense that any two solutions differ only in the choice of the coordinates t^α.

THEOREM 5.3.2. *Consider two sets of n linear differential forms, each in n variables:*

$$\alpha^i(x, dx) = \alpha^i_j(x)\,dx^j, \quad \beta^i(y, dy) = \beta^i_j(y)\,dy^j,$$

analytic at $x^i = 0$, $y^i = 0$ *respectively, and such that each of the sets* α^i, β^i *satisfies the equations*

$$d\omega^k = \gamma^k_{ij}\,\omega^i \wedge \omega^j \tag{28}$$

near the origin, where the γ*'s are certain constants. Suppose further that the n differentials* $\beta^i(0, dy)$ *are linearly independent. Then there is a unique transformation*

$$y^i = f^i(x) \quad (f^i(0) = 0),$$

analytic at $x^i = 0$, *which transforms* β^i *into* α^i.

Thus when we replace y^i, dy^j in β^h by $f^i(x)$, $\dfrac{\partial f^j}{\partial x^k}\,dx^k$ respectively, we obtain the equation

$$\beta^h\left(f(x), \frac{\partial f}{\partial x^j}\,dx^j\right) = \alpha^h(x, dx)$$

as an identity in x^i, dx^j.

Proof. Since the $\beta^i(0, dy)$ are linearly independent we can solve the equations

$$\beta^i(y, dy) = \alpha^i(x, dx) \tag{29}$$

near $y^j = 0$ for dy^i:

$$dy^i = \gamma^i_j(x, y)\,dx^j. \tag{30}$$

We now integrate these equations with the initial conditions $y^i = 0$ when $x^j = 0$. To show that they are integrable we apply exterior differentiation to (29), as in the proof of Theorem 5.2.3, and use (28). The solution has the form $y^i = f^i(x)$ and this is the required transformation since by hypothesis $\partial f^i/\partial x^j = \gamma^i_j(x, f)$, while (30) and (29) are equivalent.

COROLLARY. *If* $\alpha^i(x, dx)$ *and* $\beta^i(y, dy)$ *both satisfy* (28), *and*

$$\alpha^i(0, dx) = dx^i, \quad \beta^i(0, dy) = dy^i,$$

then there is a transformation

$$y^i = f^i(x) \quad (f^i(0) = 0),$$

analytic at 0, *with analytic inverse* $x^i = g^i(y)$ *near* 0, *which transforms* β^i *into* α^i.

This follows from the theorem, since now each of the sets α^i, β^i reduces to n linearly independent differentials at 0.

To illustrate the theorem we consider two forms ϕ and ψ in two variables x and y; we suppose that $\phi = dx$, $\psi = dy$ when $x = y = 0$.

1. Let $d\phi = 0$, $d\psi = \phi \wedge \psi$. Three solutions are

$$\begin{aligned} &(a)\ \phi = dx, && \psi = e^x\,dy,\\ &(b)\ \phi = dx, && \psi = e^{x+y^2}\,dy,\\ &(c)\ \phi = dx + 2y\,dy, && \psi = e^{x+y^2}\,dy. \end{aligned}$$

It is easily verified that these three solutions can be transformed into each other by suitable transformations.

2. $d\phi = 0$, $d\psi = \phi \wedge dx$.

Again there are several solutions:

$$\begin{aligned} &(a)\ \phi = dx, && \psi = dy,\\ &(b)\ \phi = dx + 2y\,dy, && \psi = y^2\,dx + dy. \end{aligned}$$

However, these two solutions cannot be transformed into each other. For if we denote the two coordinate systems by x, y and $\bar{x}$, $\bar{y}$, then the differentials of these coordinates must by definition satisfy the equations

$$d\bar{x} = dx + 2y\,dy, \quad d\bar{y} = y^2\,dx + dy.$$

The first equation can be integrated, giving $\bar{x} = x + y^2$, but the remaining equation is not integrable, since

$$d(d\bar{y}) = d(y^2\,dx + dy) = 2y\,dy \wedge dx \neq 0.$$

THEOREM 5.3.3 (converse of Lie's third theorem). *Let L be an abstract Lie algebra of finite dimension over R. Then there is a local Lie group $\mathfrak{B}$ whose Lie algebra is isomorphic to L. The group $\mathfrak{B}$ is determined by L to within a local analytic isomorphism.*

Proof. Let $X_1, \ldots, X_n$ be a basis of L and let

$$[X_i, X_j] = c^k_{ij} X_k \tag{31}$$

be the multiplication table in L. We shall determine n linear differential forms $\omega^1, \ldots, \omega^n$ in n variables $x^1, \ldots, x^n$ such that

$$d\omega^k = -\tfrac{1}{2} c^k_{ij}\, \omega^i \wedge \omega^j \quad (\omega^k = dx^k \text{ when } x^i = 0). \tag{32}$$

Then it is possible to determine n infinitesimal transformations X_i which satisfy (31) and reduce to $\partial/\partial x^i$ at 0, by solving the equations $\langle X_i, \omega^j \rangle = \delta^j_i$, and the theorem follows by applying Theorem 5.2.3.

In order to show that (32) is integrable we have only to verify the integrability condition of Theorem 5.3.1. This may either be done by straightforward computation or by writing again $\Omega = \omega^i X_i$. Then (32) becomes

$$d\Omega = -\tfrac{1}{2}\Omega \wedge \Omega, \tag{33}$$

and hence
$$d(d\Omega) = -\tfrac{1}{2}\{d\Omega \wedge \Omega - \Omega \wedge d\Omega\}$$
$$= \tfrac{1}{4}\{(\Omega \wedge \Omega) \wedge \Omega - \Omega \wedge (\Omega \wedge \Omega)\},$$

where we have used (33) to express $d\Omega$. Now

$$(\Omega \wedge \Omega) \wedge \Omega = -\Omega \wedge (\Omega \wedge \Omega),$$

since Ω and $\Omega \wedge \Omega$ are of degree 1 and 2 respectively; so we have

$$d(d\Omega) = \tfrac{1}{2}(\Omega \wedge \Omega) \wedge \Omega$$
$$= \tfrac{1}{2}\omega^i \wedge \omega^j \wedge \omega^k [[X_i, X_j], X_k].$$

The coefficient of $\omega^r \wedge \omega^s \wedge \omega^t$ in this expression is

$$[[X_r, X_s], X_t] + [[X_s, X_t], X_r] + [[X_t, X_r], X_s],$$

which vanishes by the Jacobi identity. Thus (32) can be solved; by Theorem 5.3.2, the solution is unique to within an analytic transformation and therefore the local group is uniquely determined to within an analytic isomorphism. This completes the proof.

Ex. Show that the equations $d\omega^1 = 0$, $d\omega^2 = -\omega^1 \wedge \omega^2$ ($\omega^1 = dx^1$, $\omega^2 = dx^2$ when $x^i = 0$) have the solution

$$\omega^1 = dx^1, \quad \omega^2 = x^2 dx^1 + dx^2.$$

CHAPTER VI

SUBGROUPS AND HOMOMORPHISMS

6.1. The canonical chart. In Chapter III we defined a mapping of the Lie algebra of a Lie group $\mathfrak{G}$ into $\mathfrak{G}$ itself, the exponential mapping. We wish to show that this mapping is analytic and invertible at the origin, but in order to do this we must of course define an analytic structure on the algebra $\Lambda(\mathfrak{G})$. Now $\Lambda(\mathfrak{G})$, as a vector space over R, has a natural analytic structure which is obtained by choosing a basis $X_1, \ldots, X_n$ and defining a chart—which covers the whole space—by taking the usual coordinates of any element X with respect to this basis. Thus if

$$X = u^i X_i, \tag{1}$$

then u^i are the coordinates of X. If we choose a different basis in $\Lambda(\mathfrak{G})$, the coordinates obtained from it will be related to the u^i by a non-singular linear transformation; thus they are certainly analytically related to the u^i, so that the analytic structure defined on $\Lambda(\mathfrak{G})$ does not depend on the choice of the basis.

Before we can prove the analyticity of the exponential mapping we need a lemma which states essentially that a power series in n variables is convergent in a neighbourhood of O (and hence defines a function which is analytic at O) provided that in each direction there is a finite segment starting at O, on which the power series is convergent.

LEMMA 6.1.1. *Let $p(x)$ be a power series in $x^1, \ldots, x^n$ and suppose that to each point λ on the unit sphere*

$$S_n: \Sigma |\lambda^i|^2 = 1$$

there corresponds a positive constant $\gamma(\lambda)$ such that the series $p(x)$ converges for $x^i = t\lambda^i$, where t is any real number satisfying

$$|t| < \gamma(\lambda).$$

Then the function defined by $p(x)$ is analytic at O, i.e. there exists a constant $\gamma > 0$ such that $p(x)$ converges for all x^i satisfying

$$\Sigma |x^i|^2 < \gamma^2.$$

Proof. Let $\lambda_0 = (\lambda_0^i)$ be any point on S_n. We first show that if $p(x)$ converges for all $x^i = t\lambda_0^i$ such that $|t| < \gamma(\lambda_0)$, then $p(x)$ converges for $x^i = t\lambda^i$ whenever $|t| < \frac{1}{2}\gamma(\lambda_0)$ and λ belongs to a certain neighbourhood of λ_0.

For $p(x)$ converges for $x^i = t\lambda_1^i$, where $|t| < \frac{1}{2}\gamma(\lambda_0)$ and $\lambda_1^i = 2\lambda_0^i$. Since it is a power series in the λ_1^i, it also converges for any λ^i satisfying $|\lambda^i| < 2|\lambda_0^i|$, and the region of S_n defined by

$$|\lambda^i| < 2|\lambda_0^i| \quad (i = 1, \ldots, n),$$
$$\Sigma|\lambda^i|^2 = 1,$$

is clearly a neighbourhood of λ_0.

The lemma now follows by the compactness of S_n: We have shown that each point λ_0 of S_n has a neighbourhood $U(\lambda_0)$—which may be taken to be open—on which the radius of convergence in t has a positive lower bound, viz. $\frac{1}{2}\gamma(\lambda_0)$. The open sets $U(\lambda_0)$ cover S_n, by the compactness we can choose a finite subcovering, $U_1, \ldots, U_k$ say, and the minimum, γ say, of the lower bounds on these sets U_i is a universal lower bound. Thus $p(x)$ converges for $x^i = t\lambda^i$, where $\Sigma|\lambda^i|^2 = 1$ and $|t| < \gamma$; in other words, for all x^i satisfying $\Sigma|x^i|^2 < \gamma^2$. This is what we had to prove.

If we denote by $c(\lambda)$ the exact radius of convergence (in t) in the direction λ, so that $c(\lambda)$ is the least upper bound of the possible values of $\gamma(\lambda)$ in the lemma, then the first part of the proof shows essentially that this function $c(\lambda)$ is lower semi-continuous on S_n, and the second part amounts to showing that on a compact set such a function attains its greatest lower bound.† Later on we shall meet another application of this second part. For reference we state the relevant part of Lemma 6.1.1 as follows:

LEMMA 6.1.2. *Let $\gamma(\lambda)$ be a real function on the unit sphere S_n: $\Sigma|\lambda^i|^2 = 1$. If each point of S_n has a neighbourhood on which $\gamma(\lambda)$ has a positive lower bound, then $\gamma(\lambda)$ has a positive lower bound as λ varies over the whole of S_n.*

THEOREM 6.1.3. *Let $\mathfrak{G}$ be any Lie group, $\Lambda(\mathfrak{G})$ its Lie algebra; then the mapping $X \to \exp X (X \in \Lambda(\mathfrak{G}))$ is analytic and invertible at O.*

† Cf. Bourbaki[2], p. 112.

Proof. Let (u) be a chart in $\Lambda(\mathfrak{G})$ defined by a basis

$$X_i \, (i=1, \ldots, n)$$

f $\Lambda(\mathfrak{G})$, as in (1), and let (x) be a chart at e in $\mathfrak{G}$. The tangent ectors $(X_i)_e$ form a basis of $\mathfrak{L}_e(\mathfrak{G})$, and so do the tangent vectors $/\partial x^i$ at e. Therefore they are related by equations

$$(X_i)_e = \alpha_i^j \frac{\partial}{\partial x^j},$$

where the α_i^j are constants with a non-vanishing determinant. Let us calculate the coordinates of the point $\exp u^i X_i$ of $\mathfrak{G}$. They are given by

$$g^i(u) = [\exp u^h X_h x^i]_{x=e}.$$

If we expand the right-hand side, we obtain a power series in the u^i, which is convergent for all points $u = (u^1, \ldots, u^n)$ which, in any given direction, are less than a certain positive distance (depending on the given direction) from the origin (cf. 3.6 (31)). It follows by Lemma 6.1.1 that $g^i(u)$ is analytic at O, whence it follows that the exponential mapping is analytic at O.

Now we have

$$g^i(u) = [(1 + u^h X_h) x^i + o(\eta)]_{x=e},$$

where $o(\eta)$ is a term of smaller order than $\eta = (\Sigma \mid u^i \mid^2)^{\frac{1}{2}}$; hence

$$\left[\frac{\partial g^i}{\partial u^j}\right]_{x=e} = [X_j x^i]_{x=e} = \alpha_j^i.$$

Since $\dim \mathfrak{G} = \dim \Lambda(\mathfrak{G})$ and $\det(\alpha_i^j) \neq 0$, the hypotheses of Theorem 1.8.3 are satisfied, and so $\exp X$ is invertible at O, as we wished to show.

As a consequence of Theorem 6.1.3 we may use the exponential mapping to transfer any chart at O in $\Lambda(\mathfrak{G})$ to $\mathfrak{G}$, giving a chart at e, and vice versa. But whereas all the charts at e in $\mathfrak{G}$ are *a priori* on an equal footing, there is a privileged class of charts in $\Lambda(\mathfrak{G})$, namely, those which are obtained by taking the coordinates with respect to a basis of $\Lambda(\mathfrak{G})$ as in (1).† The corresponding charts in $\mathfrak{G}$ and their coordinates are called *canonical*. Thus a chart at e is canonical if, for a suitable choice of

† Of course an admissible chart at O in $\Lambda(\mathfrak{G})$ is, in general, not of this special form, since there are curvilinear coordinate systems in $\Lambda(\mathfrak{G})$ as well as Cartesian ones.

the basis X_i of $\Lambda(\mathfrak{G})$, the coordinates of the point $\exp u^i X_i$ are $u^1, \ldots, u^n$. In symbols,

$$(\exp u^i X_i)^j = u^j. \tag{2}$$

By considering the relation between different bases of $\Lambda(\mathfrak{G})$, we see that the coordinates of two canonical charts are related by a non-singular linear transformation, and conversely, any chart at e, whose coordinates are obtained from canonical coordinates by a non-singular linear transformation, is again canonical.

Canonical charts can often be used to simplify proofs, and they are sometimes essential. Occasionally a different type of chart is defined in $\mathfrak{G}$ by taking $u^1, \ldots, u^n$ to be the coordinates of the point

$$\exp u^1 X_1 . \exp u^2 X_2 \ldots \exp u^n X_n. \tag{3}$$

It is easily verified that this does in fact define a chart at e. The coordinates defined by (3) are sometimes called canonical coordinates of the second kind; then the coordinates (2) are called canonical of the first kind. We shall not use this terminology.

Another type of chart is obtained by choosing an integer ν in the range $1 \leqslant \nu \leqslant n$ and taking $u^1, \ldots, u^n$ to be the coordinates of the point

$$\exp (u^1 X_1 + \ldots + u^\nu X_\nu) . \exp (u^{\nu+1} X_{\nu+1} + \ldots + u^n X_n).\dagger \tag{4}$$

We shall have occasion to use this type of chart later. The task of verifying that (4) actually defines a chart is again left to the reader.

Since the exponential mapping is invertible at O, the image of $\Lambda(\mathfrak{G})$ under the mapping is a neighbourhood of e, in other words, there is a nucleus V of $\mathfrak{G}$ such that every $x \in V$ is of the form $\exp X$. If $X, Y \in \Lambda(\mathfrak{G})$ and t is real, then the elements

$$\exp (tX) . \exp (tY) \quad \text{and} \quad (\exp (tX), \exp (tY))\ddagger$$

belong to V for sufficiently small t, and hence can be expressed in the form $\exp Z$. The form which Z takes is described by

† No summation.

‡ For brevity we use the notation $(x, y) = x^{-1} y^{-1} x y$ for elements of a group.

THEOREM 6.1.4. *Let $X, Y \in \Lambda(\mathfrak{G})$ and let t be a real parameter tending to zero. Then*

(i) $\exp(tX).\exp(tY) = \exp\{t(X+Y)+O(t^2)\}$,

(ii) $\exp(tX).\exp(t^2Y) = \exp\{tX+t^2Y+O(t^3)\}$,

(iii) $(\exp(tX), \exp(tY)) = \exp\{t^2[X,Y]+O(t^3)\}$,

(iv) $\exp t(X+Y) = \exp(tX).\exp(tY).\exp O(t^2)$,

(v) $\exp t^2[X,Y] = (\exp(tX), \exp(tY)).\exp O(t^3)$.

Proof. (i) For sufficiently small t we have

$$\exp(tX)\exp(tY) = \exp Z(t), \tag{5}$$

say, where $Z(t) \in \Lambda(\mathfrak{G})$. Since the exponential mapping is analytic and invertible at O, $Z(t)$ is an analytic function of t at 0; for $t=0$ it vanishes, by (5), and so we have $Z = tZ_1 + O(t^2)$, say. Substituting in (5) and expanding, we find

$$(1+tX+\ldots)(1+tY+\ldots) = 1+tZ_1+\ldots,$$

where the terms which are $O(t^2)$ are indicated by dots. Hence

$$1+t(X+Y)+\ldots = 1+tZ_1+\ldots;$$

if we subtract 1 from both sides, divide by t and let $t \to 0$, we find that $Z_1 = X+Y$, whence $Z = t(X+Y)+O(t^2)$, as asserted. The proof of (ii) is similar and may be omitted. To prove (iii) we have

$$(\exp(tX), \exp(tY)) = \exp Z(t),$$

where $Z(t)$ is again an analytic function of t which vanishes for $t=0$. When we expand the left-hand side and indicate the $O(t^3)$ terms by dots, we get

$$(1-tX+\tfrac{1}{2}t^2X^2-\ldots)(1-tY+\tfrac{1}{2}t^2Y^2-\ldots)(1+tX+\tfrac{1}{2}t^2X^2+\ldots) \\ \times(1+tY+\tfrac{1}{2}t^2Y^2+\ldots) = 1+t^2(XY-YX)+\ldots.$$

If we put $Z(t) = tZ_1+t^2Z_2+\ldots$ and with this value expand the right-hand side, we obtain

$$1+tZ_1+t^2Z_2+\tfrac{1}{2}(tZ_1+t^2Z_2)^2+\ldots,$$

which equals $$1+tZ_1+t^2(Z_2+\tfrac{1}{2}Z_1^2)+\ldots.$$

A comparison shows that $Z_1 = 0$, $Z_2 = XY - YX = [X, Y]$, hence $Z = t^2[X, Y] + O(t^3)$ and (iii) follows. Now (iv) and (v) follow by repeated application of (i) and (iii). Thus

$$\begin{aligned}\exp(-tY)\exp(-tX)&\exp t(X+Y)\\ &= \exp\{-t(X+Y)+O(t^2)\}\exp t(X+Y)\\ &= \exp O(t^2),\end{aligned}$$

from which (iv) follows by multiplying on the left by

$$\exp(tX)\exp(tY).$$

Similarly,

$$\begin{aligned}(\exp(tX),\ \exp(tY))^{-1}&\exp t^2[X,Y]\\ &= \exp\{-t^2[X,Y]+O(t^3)\}\exp t^2[X,Y]\\ &= \exp O(t^3),\end{aligned}$$

and this leads to (v) when we multiply on the left by

$$(\exp(tX),\ \exp(tY)).$$

Ex. 1. In a canonical chart (defined by (2)),

$$(x^{-1})^i = -x^i,$$

for any point x in the appropriate nucleus of $\mathfrak{G}$.

Ex. 2. Find canonical coordinates for the affine group (defined in 3.3, Ex. 1).

Ex. 3. Show that

$$\left\{\exp\frac{X}{n}\exp\frac{Y}{n}\right\}^n \to \exp(X+Y)$$

and

$$\left\{\exp\frac{X}{n},\ \exp\frac{Y}{n}\right\}^{n^2} \to \exp[X,Y],$$

as $n \to \infty$.

6.2. A characterization of canonical charts. One of the main applications of canonical charts is the proof that every continuous homomorphism of one (real) Lie group into another is analytic.† This is based on the following theorem which gives a characterization of canonical charts depending only on the structure of $\mathfrak{G}$ as a locally Euclidean group:

† Cf. Ex. 2 at the end of 6.3.

THEOREM 6.2.1. *In a local Lie group with coordinates $x^1, \ldots, x^n$ defined in a nucleus V, the following statements are equivalent:*

(*a*) *the x^i define a canonical chart in V;*

(*b*) *if $\psi^i_j(x)$ are the transformation functions, then*

$$\psi^i_j(x)\, x^j = x^i \quad (x \in V); \tag{6}$$

(*c*) *for any constants $\lambda^1, \ldots, \lambda^n$, the mapping $t \to g(t)$, where $g^i(t) = \lambda^i t$, is a local homomorphism of $\mathfrak{R}$ into V.*

Moreover, when these conditions are satisfied, every 1-parameter subgroup can be written as $g^i(t) = \lambda^i t$ for some constants $\lambda^1, \ldots, \lambda^n$ not all zero.

Proof. 1. (*b*) implies (*c*). If in (6) we replace x^i by $\lambda^i t$ and divide by t, we find

$$\lambda^i = \lambda^j \psi^i_j(\lambda t).$$

Hence the functions $x^i = \lambda^i t$ satisfy the equations

$$\frac{dx^i}{dt} = \lambda^j \psi^i_j(x), \tag{7}$$

with the initial conditions $x^i = 0$ when $t = 0$. By Theorem 3.5.2, the mapping $t \to (\lambda^1 t, \ldots, \lambda^n t)$ defines a local homomorphism of $\mathfrak{R}$ into a 1-parameter local subgroup of V, and this is analytic because $\lambda^i t$ is analytic in t.

2. (*c*) implies (*b*). Again by Theorem 3.5.2, if $t \to (\lambda^i t)$ is a local analytic homomorphism then the functions $x^i = \lambda^i t$ satisfy (7), i.e.

$$\lambda^i = \lambda^j \psi^i_j(\lambda t).$$

Putting $t = 1$ and replacing λ^i by x^i we obtain (6); thus (*b*) holds.

3. (*a*) implies (*c*). Condition (*a*) states that the coordinates of $\exp u^i X_i$ are u^j; thus we have to show that the mapping

$$t \to \exp \lambda^i t X_i$$

is a local homomorphism. Write $X = \lambda^i X_i$, then the mapping becomes $t \to \exp tX$, and this is a local homomorphism, since

$$\exp t_1 X . \exp t_2 X = \exp (t_1 + t_2) X$$

for sufficiently small t_1, t_2 (cf. 3.7).

4. (*c*) implies (*a*). If (*c*) holds, then for any constants λ^i, not all 0, the mapping $t \to g(t)$, where $g^i(t) = \lambda^i t$, is a local homomorphism, whence $g(t_1)\, g(t_2) = g(t_1 + t_2)$ for all t_1, t_2 near 0.

Therefore the elements $g(t)$ form a 1-parameter local subgroup U, say, of V. Let X_i be a basis of the Lie algebra of V; we may suppose this basis so chosen that $(X_i)_e = \partial/\partial x^i$. The canonical chart with respect to this basis is defined by

$$(\exp u^i X_i)^j = u^j.$$

We shall prove that on U we have $u^i(t) = g^i(t)$. Since there is some nucleus in V, each of whose points lies on a 1-parameter subgroup, this will prove that $x^i = u^i$, i.e. the given coordinates are in fact canonical.

In some nucleus the element $g(t)$ may be expressed as an image under the exponential mapping: $g(t) = \exp A(t)$, where

$$A(t) = u^i X_i$$

with coordinates u^i which depend analytically on t: $u^i = u^i(t)$. Since U is Abelian, we have, by Theorem 6.1.4,

$$e = (\exp sA(t_1), \exp sA(t_2)) = \exp\{s^2[A(t_1), A(t_2)] + O(s^3)\}$$

for small s. Therefore

$$s^2[A(t_1), A(t_2)] + O(s^3) = 0.$$

Dividing by s^2 and letting $s \to 0$, we obtain $[A(t_1), A(t_2)] = 0$, whence $\exp A(t_1) \exp A(t_2) = \exp(A(t_1) + A(t_2))$. But we also have, from the definition of $A(t)$,

$$\exp A(t_1) \exp A(t_2) = \exp A(t_1 + t_2).$$

Hence $A(t_1 + t_2) = A(t_1) + A(t_2)$, i.e.

$$\{u^i(t_1 + t_2) - u^i(t_1) - u^i(t_2)\} X_i = 0,$$

whence
$$u^i(t_1 + t_2) = u^i(t_1) + u^i(t_2).$$

Now u^i is an analytic function of t, and therefore $u^i(t) = \mu^i t$, where the μ^i are constants which may be calculated as follows: By the definition of $g^i(t)$,

$$\lambda^k t = g^k(t) = [\exp \mu^i t X_i x^k]_{x=e} = [x^k + \mu^k t + O(t^2)]_{x=e}$$
$$= \mu^k t + O(t^2).$$

Hence $\lambda^k = \mu^k$, and so $u^i(t) = \lambda^i t = g^i(t)$, as we wished to show.

5. Finally, every 1-parameter subgroup is determined completely by its tangent vector at e, λ^i say, where the λ^i are not

all 0. If (c) holds, then $g^i(t) = \lambda^i t$ is a 1-parameter subgroup with this tangent vector, therefore every 1-parameter subgroup can be expressed in this form. This completes the proof.

We note that by the equivalence of (a) and (c), the canonical coordinate defined in 2.9 is a special case of the present concept. The importance of Theorem 6.2.1 lies in the fact that it links a particular type of chart at e—the canonical charts—with entities depending only on the group, namely the 1-parameter subgroups.

Ex. Calculate the transformation functions for the canonical chart found in 6.1, Ex. 2, and verify (6) in this case.

6.3. Continuous homomorphisms. We can now show that continuous homomorphisms between real Lie groups are analytic. We remark that it is enough to prove the analyticity at a single point, for if θ is a homomorphism of one Lie group $\mathfrak{G}$ into another, which is analytic at the unit element e of $\mathfrak{G}$, then since $\theta(ax) = \theta(a)\,\theta(x)$, it follows that $\theta(ax)$ is analytic at $x = e$, and hence θ is analytic at a. For convenience the 1-dimensional case is treated first.

THEOREM 6.3.1. *Every continuous homomorphism of a 1-dimensional Lie group $\mathfrak{G}$ into a Lie group $\mathfrak{H}$ is analytic.*

Proof. Let the homomorphism be $t \to g(t)$; we have to prove that in terms of some chart, the coordinates $g^i(t)$ are analytic functions of t. If we choose the coordinate in $\mathfrak{G}$ to be canonical, then

$$g(t_1)\,g(t_2) = g(t_1 + t_2); \tag{8}$$

if, further, we take the coordinates in $\mathfrak{H}$ also to be canonical, then

$$g(t) = \exp g^i(t)\,X_i,$$

and hence, since by (8) the elements $g(t)$ commute,

$$g(t_1)\,g(t_2) = \exp g^i(t_1)\,X_i \,.\, \exp g^i(t_2)\,X_i = \exp\{g^i(t_1) + g^i(t_2)\}\,X_i.$$

Similarly, $g(t_1 + t_2) = \exp g^i(t_1 + t_2)\,X_i$, and therefore, by (8),

$$g^i(t_1 + t_2) = g^i(t_1) + g^i(t_2).$$

The theorem will follow if we can show that any continuous function satisfying this equation is linear. For since the g^i are continuous, they must then be linear and hence analytic.

Let f then be a continuous function of a real variable satisfying

$$f(s + t) = f(s) + f(t) \tag{9}$$

for s and t in some neighbourhood of 0. We can extend the definition of f so that it is defined for all real s and satisfies (9), by Lemma 2.9.2, and the extended function is continuous, because f is continuous at 0 (Theorem 2.3.4). Let $f(1)=\lambda$, then $f(n)=n\lambda$ and $mf(n/m)=f(n)=n\lambda$, whence $f(n/m)=n\lambda/m$ $(m\neq 0)$. Thus we have $f(t)=t\lambda$ for all rational t, and by continuity, for all real t, in other words, f is linear, as we wished to show.

We can now deduce the general proposition by making use of the fact that every Lie group has a nucleus which can be covered by 1-parameter subgroups.

THEOREM 6.3.2. *Any continuous homomorphism of a Lie group $\mathfrak{G}$ into a Lie group $\mathfrak{H}$ is analytic.*

Proof. Let θ be the given homomorphism; further, let $\mathfrak{g}$ and $\mathfrak{h}$ be the Lie algebras of $\mathfrak{G}$ and $\mathfrak{H}$ respectively, and denote by $X_1, \ldots, X_n$ a basis of $\mathfrak{g}$. Then for each $i=1, \ldots, n$, $\{\exp sX_i\}$ is a 1-parameter subgroup of $\mathfrak{G}$, and hence $\{(\exp sX_i)^\theta\}$ is either a 1-parameter subgroup of $\mathfrak{H}$, or the unit element of $\mathfrak{H}$. In either case we may, by Theorem 6.3.1, write the image as $\exp sY_i$, where $Y_i \in \mathfrak{h}$. Now there is a nucleus U of $\mathfrak{G}$ in which every element can be expressed in the form

$$\exp u^1X_1 . \exp u^2X_2 \ldots \exp u^nX_n.$$

For any such element we have

$$\begin{aligned}(\exp u^1X_1 \ldots \exp u^nX_n)^\theta &= (\exp u^1X_1)^\theta \ldots (\exp u^nX_n)^\theta \\ &= \exp u^1Y_1 \ldots \exp u^nY_n.\end{aligned}$$

The expression on the right is analytic in $u^1, \ldots, u^n$, since its coordinates are analytic functions of the u's; therefore θ is analytic at the unit element of $\mathfrak{G}$, and hence everywhere. This completes the proof.

As a corollary we have the result already mentioned in Chapter II:

THEOREM 6.3.3. *Two Lie groups with the same underlying topological group coincide.*

For under the given hypothesis the identity mapping on the underlying space is a topological isomorphism of the one group onto the other, and hence, by Theorem 6.3.2, an analytic isonorphism. This means that the two analytic structures coincide nd the theorem follows.

The same argument shows that even a *local* isomorphism between Lie groups, which is continuous, is necessarily analytic. Moreover, it is enough to assume continuity in one direction; as we shall need this fact later, we state it as a theorem:

THEOREM 6.3.4. *If* $\mathfrak{G}$ *and* $\mathfrak{H}$ *are Lie groups of which* $\mathfrak{G}$ *is connected and* θ *is a continuous homomorphism of* $\mathfrak{G}$ *into* $\mathfrak{H}$ *which is one-one on a nucleus of* $\mathfrak{G}$, *then* θ *is a local analytic isomorphism between* $\mathfrak{G}$ *and an analytic subgroup of* $\mathfrak{H}$.

For θ defines a continuous one-one mapping of a nucleus V of $\mathfrak{G}$ into $\mathfrak{H}$. By restricting V suitably we may take it to be compact, and since any continuous one-one mapping of a compact space is a homeomorphism, it follows that the restriction $\theta \mid V$ is a topological isomorphism. We can now deduce, as in Theorem 6.3.2, that $\theta \mid V$, as well as its inverse, is analytic. The interior of the image V^θ is a local Lie group which generates an analytic subgroup of $\mathfrak{H}$; now θ provides a local analytic isomorphism between $\mathfrak{G}$ and this group, and so the proof is complete.

As an application of Theorem 6.3.2, we shall show how to simplify the definition of an analytic subgroup. By Theorem 2.8.1, the definition may be stated thus: $\mathfrak{H}$ is an analytic subgroup of $\mathfrak{G}$ if

(i) $\mathfrak{H}$ is a subgroup of $\mathfrak{G}$,

(ii) $\mathfrak{H}$ is a Lie group,

(iii) given a chart at e in $\mathfrak{G}$, we can select coordinates from it whose restrictions to $\mathfrak{H}$ define a chart at e in $\mathfrak{H}$.

Conditions (i) and (ii) are of course essential, and examples show that (iii) cannot be omitted altogether (cf. Ex. 3 below). But we shall see that it is enough to assume that the identity mapping of $\mathfrak{H}$ into $\mathfrak{G}$ is continuous; moreover, the analytic structure is then uniquely determined.

THEOREM 6.3.5. *Let* $\mathfrak{G}$ *be a Lie group and* $\mathfrak{H}$ *a subset of* $\mathfrak{G}$ *such that*

(i) $\mathfrak{H}$ *is a subgroup of* $\mathfrak{G}$ *(qua abstract group),*

(ii) $\mathfrak{H}$ *is a Lie group,*

(iii) *the identity mapping of* $\mathfrak{H}$ *into* $\mathfrak{G}$ *is continuous.*

Then $\mathfrak{H}$ *can be defined in just one way as an analytic subgroup of* $\mathfrak{G}$.

Proof. Denote by ι the identity mapping of $\mathfrak{H}$ into $\mathfrak{G}$. Since $\mathfrak{H}$ is a subgroup of $\mathfrak{G}$, ι is an isomorphism of $\mathfrak{H}$ into $\mathfrak{G}$, and being continuous, it must be analytic, by Theorem 6.3.2.

Let $(x^1, \ldots, x^n)$, $(u^1, \ldots, u^\nu)$ be charts at e in $\mathfrak{G}$, $\mathfrak{H}$ respectively. Since ι is analytic, the functions $\bar{x}^i = x^i \mid \mathfrak{H}$ are analytic. Let us consider, on $\mathfrak{H}$, the Maurer-Cartan forms ϖ^i, θ^α which reduce to $d\bar{x}^i$, du^α respectively at e. Since the θ^α form a basis for the linear MC-forms on $\mathfrak{H}$, we have†

$$\varpi^i = c^i_\alpha \theta^\alpha, \tag{10}$$

where the c^i_α are constants; taking this equation at e, we see that $d\bar{x}^i = c^i_\alpha du^\alpha$, whence $c^i_\alpha = (\partial \bar{x}^i / \partial u^\alpha)_e$. Let ρ be the rank of the matrix (c^i_α); then $\rho \leqslant \nu$, where ν is the dimension of $\mathfrak{H}$. If we can show that $\rho = \nu$, then it follows from Theorem 2.8.1 that $\mathfrak{H}$ is an analytic subgroup of $\mathfrak{G}$, so let us suppose that $\rho < \nu$. The matrix $(\partial \bar{x}^i / \partial u^\alpha)$ is of the same rank as (c^i_α), viz. ρ, wherever it is defined, because at every point the $d\bar{x}^i$ are related to the ϖ^j by a non-singular transformation, and likewise for the du^α and θ^β. We may suppose the x^i and u^α so numbered that the minor involving $\bar{x}^1, \ldots, \bar{x}^\rho$ and $u^1, \ldots, u^\rho$ is $\neq 0$ at e. Then we can solve for $u^1, \ldots, u^\rho$ in terms of $\bar{x}^1, \ldots, \bar{x}^\rho$ near e, and hence we may replace $u^1, \ldots, u^\rho$ by $\bar{x}^1, \ldots, \bar{x}^\rho$ respectively, in the chart (u). Calling these co-ordinates again $u^1, \ldots, u^\rho$, we now have charts (x), (u) as before, and in addition
$$\bar{x}^1 = u^1, \quad \ldots, \quad \bar{x}^\rho = u^\rho.$$

Since the new $n \times \nu$ matrix

$$\left(\frac{\partial \bar{x}^i}{\partial u^\alpha}\right) = \begin{pmatrix} I & 0 \\ * & * \end{pmatrix}$$

is again of rank ρ, it follows that $\partial \bar{x}^i / \partial u^\beta = 0$ for $\beta > \rho$ throughout some nucleus of $\mathfrak{H}$. Therefore the $\bar{x}^i$ are independent of

$$u^{\rho+1}, \quad \ldots, \quad u^\nu;$$

i.e. the values of $\bar{x}^1, \ldots, \bar{x}^n$ are completely determined by $u^1, \ldots, u^\rho$. This contradicts the hypothesis that $(u^1, \ldots, u^\nu)$ defines a chart on a nucleus of $\mathfrak{H}$, unless $\rho = \nu$.

† We use Greek suffixes, with the summation convention, for the range $1, \ldots, \nu$.

Finally, the analytic subgroup structure of $\mathfrak{H}$ is unique, since the analytic structures of both $\mathfrak{H}$ and $\mathfrak{G}$ are given. This completes the proof.

We note without proof that a given abstract subgroup of a Lie group $\mathfrak{G}$ can be defined as an analytic subgroup of $\mathfrak{G}$ in at most one way. In other words, the Lie group structure of $\mathfrak{H}$, in Theorem 6.3.5, is already determined by that of $\mathfrak{G}$, together with the condition (iii). Of course a Lie group structure does not in general exist for subgroups of $\mathfrak{G}$ which are topological groups and satisfy (iii); consider, for example, the set of rational numbers as a subgroup of $\mathfrak{R}$ with the topology induced by $\mathfrak{R}$. However, if H is a subgroup satisfying (iii) and H is *locally compact* (i.e. if there is a compact nucleus of H), then H can be proved to be a Lie group (cf. Chevalley [5]).

Ex. 1. If two Lie groups are locally topologically isomorphic, then they have isomorphic Lie algebras.

Ex. 2. Let C be the field of complex numbers. By replacing R by C in the definition of an analytic manifold, we obtain a *complex analytic manifold*. A complex Lie group is then defined as a group which is also a complex analytic manifold such that the multiplication is analytic. In particular C itself is a 1-dimensional complex Lie group with respect to addition, denoted by $\mathfrak{C}$. Show that the analogue of Theorem 6.3.3 is false in the case of the group $\mathfrak{C}$.

Ex. 3. By considering R as a vector space over the field of rational numbers (using a Hamel basis) show that there are distinct Lie group structures on R, and hence that a Lie group which is subgroup of another Lie group, is not necessarily an analytic subgroup.

6.4. Analytic subgroups and their Lie algebras. Our next objective is to establish a correspondence between the subgroups of a Lie group $\mathfrak{G}$ and the subalgebras of the Lie algebra $\mathfrak{g}$ of $\mathfrak{G}$. We saw already in Chapter III that to every analytic subgroup of $\mathfrak{G}$ there corresponds a subalgebra of $\mathfrak{g}$, and we shall now show that in this correspondence there is an analytic subgroup for every subalgebra. The subgroup will, in general, not be unique, since any analytic subgroup and its identity

I

component correspond to the same Lie algebra, but at least the subgroup will be determined to within local isomorphism. Since any Lie group is locally isomorphic to its identity component, we may restrict our attention to connected groups, and here we shall find that there is a one-one correspondence.†

The dimension of the whole group will be denoted by n, as before, and the dimension of the analytic subgroup considered by ν, where $0 \leqslant \nu \leqslant n$, though sometimes we shall suppose $\nu \neq 0$ or $\nu \neq n$. In the formulae suffixes in the range $1, \ldots, n$ will again be denoted by $i, j, k, \ldots$, while suffixes in the partial ranges $1, \ldots, \nu$ and $\nu+1, \ldots, n$ will be denoted by $\alpha, \beta, \ldots$ and $\alpha', \beta', \ldots$ respectively. A repeated suffix is understood to be summed over the appropriate range, unless otherwise stated.

THEOREM 6.4.1. *Let $\mathfrak{G}$ be a Lie group and $\mathfrak{g}$ its Lie algebra. If $\mathfrak{h}$ is any subalgebra of $\mathfrak{g}$, then the subgroup $\mathfrak{H}$ of $\mathfrak{G}$ generated by the elements* $\exp Y$ $(Y \in \mathfrak{h})$ *is a connected analytic subgroup of $\mathfrak{G}$, whose Lie algebra is $\mathfrak{h}$.*

Proof. Let the dimensions of $\mathfrak{g}$ and $\mathfrak{h}$ be n and ν respectively. If $\nu = 0$, then $\mathfrak{h} = 0$ and $\mathfrak{H}$ consists of e alone. The theorem is then trivially true and so we may suppose $1 \leqslant \nu \leqslant n$. Let $X_1, \ldots, X_n$ be a basis of $\mathfrak{g}$, chosen so that $X_1, \ldots, X_\nu$ is a basis of $\mathfrak{h}$, and let $\omega^1, \ldots, \omega^n$ be the dual basis of MC-forms. Now select a chart $(x^1, \ldots, x^n)$ at e, defined on a nucleus V of $\mathfrak{G}$, say. By applying a linear transformation to the x's if necessary we may suppose that $x^i = 0$ at e, and

$$\omega^i = dx^i \quad \text{at } e. \tag{11}$$

From the results of Chapter IV we know that the ω^i satisfy the equations

$$d\omega^k = -\tfrac{1}{2} c^k_{ij} \omega^i \wedge \omega^j, \tag{12}$$

where the c^k_{ij} are the structure constants for the basis X_i. Our aim is to make a change of coordinates at e in such a way that a certain ν-dimensional subset of the nucleus V is a local subgroup with the Lie algebra $\mathfrak{h}$. To this end we consider the equations

$$\left.\begin{aligned} d\theta^\gamma &= -\tfrac{1}{2} c^\gamma_{\alpha\beta} \theta^\alpha \wedge \theta^\beta, \\ \theta^\gamma &= du^\gamma \quad \text{when } u^\alpha = 0. \end{aligned}\right\} \tag{13}$$

† Cf. the case of 1-dimensional subgroups, considered in Chapter III.

This is a system of equations for the differential forms $\theta^1, \ldots, \theta^\nu$, and the equations are integrable; the condition of Theorem 5.3.1 is satisfied because $\mathfrak{h}$ is a Lie algebra. Let the solution of (13) be

$$\theta^\alpha = \varpi^\alpha(u, du),$$

and define $\varpi^{\alpha'} = 0$ for $\alpha' > \nu$. Then the forms ϖ^k satisfy (12): If $k = \gamma \leqslant \nu$, equation (12) reduces to

$$d\varpi^\gamma = -\tfrac{1}{2}c^\gamma_{\alpha\beta}\varpi^\alpha \wedge \varpi^\beta,$$

since $\varpi^{\alpha'} = 0$ for $\alpha' > \nu$. If $k = \gamma' > \nu$, then (12) becomes

$$d\varpi^{\gamma'} = -\tfrac{1}{2}c^{\gamma'}_{\alpha\beta}\varpi^\alpha \wedge \varpi^\beta,$$

and here the right-hand side vanishes because $\mathfrak{h}$ is a subalgebra.

Thus the forms ω^i, ϖ^i satisfy the same equations (12), and, moreover, the ω^i are linearly independent at e, by (11). Hence, applying Theorem 5.3.2, we may integrate the equations

$$\omega^i(x, dx) = \varpi^i(u, du) \tag{14}$$

in the form $$x^i = f^i(u^1, \ldots, u^\nu) \quad (f^i(0) = 0). \tag{15}$$

Now the equations $\varpi^i(w, dw) = \varpi^i(v, dv)$ are integrable, by the criterion of Theorem 5.2.1. Let

$$w^\alpha = \chi^\alpha(u, v) \tag{16}$$

be that solution which reduces to u^α, when $v^\alpha = 0$. We shall show that the ν-dimensional subset defined by (15) in some nucleus is a local subgroup of $\mathfrak{G}$ by proving: If x and y are any two points of this set, given by $x^i = f^i(u)$, $y^i = f^i(v)$, and if $z^i = (xy)^i = \phi^i(x, y)$, $w^\alpha = \chi^\alpha(u, v)$, then

$$z^i = f^i(w).$$

This will prove the set (15) to be a local subgroup of $\mathfrak{G}$, and (16) its composition functions.

By the definition of v,

$$\omega^i(y, dy) = \varpi^i(v, dv),$$

and by the definitions of z and w respectively,

$$\omega^i(z, dz) = \omega^i(y, dy),$$
$$\varpi^i(w, dw) = \varpi^i(v, dv);$$

hence $$\omega^i(z, dz) = \varpi^i(w, dw). \tag{17}$$

When $u^\alpha = v^\alpha = 0$, w^α reduces to 0, and then $z^i = \phi^i(0, 0) = ($ therefore, by the uniqueness of the solution (15), the integratio of (17) gives

$$z^i = f^i(w).$$

Thus (15) and (16) define a local Lie group, of dimension ν, whicl is a subgroup of $\mathfrak{G}$. Let $\mathfrak{H}$ be the subgroup of $\mathfrak{G}$ generated by thi local group, then $\mathfrak{H}$ is a connected ν-dimensional analytic sub group of $\mathfrak{G}$, by Theorem 2.8.2.

Now the two sets of forms ω^i and ϖ^i are both left-invarian and on $\mathfrak{H}$ they are equal at e (by (14)). Since we can go to any point u of $\mathfrak{H}$ by the left translation λ_u, the forms are equal over the whole of $\mathfrak{H}$. The elements $X_1, \ldots, X_\nu$ of $\mathfrak{g}$ satisfy the equations

$$\langle X_\alpha, \omega^\beta \rangle = \delta^\beta_\alpha;$$

hence we have also
$$\langle X_\alpha, \varpi^\beta \rangle = \delta^\beta_\alpha, \tag{18}$$

which shows that the X_α just form a basis of the Lie algebra of $\mathfrak{H}$. For the ϖ^β form a basis of the algebra of MC-forms on $\mathfrak{H}$, and the X_α are uniquely determined by (18). Therefore $\mathfrak{h}$ is the Lie algebra of $\mathfrak{H}$. By Theorem 3.5.1 and the definition of the exponential mapping it follows that $\exp Y \in \mathfrak{H}$ for all $Y \in \mathfrak{h}$. Hence, if $\mathfrak{H}_1$ is the subgroup of $\mathfrak{G}$ generated by the elements $\exp Y (Y \in \mathfrak{h})$, then $\mathfrak{H}_1 \subseteq \mathfrak{H}$. But $\mathfrak{H}_1$ has a nucleus which is a submanifold of dimension ν, viz. the set of all points $\exp u^\alpha X_\alpha$. Hence $\mathfrak{H}_1$ is a ν-dimensional analytic subgroup of $\mathfrak{H}$, and because $\mathfrak{H}$ is connected and of dimension ν, it follows that $\mathfrak{H}_1 = \mathfrak{H}$ and the proof is complete.

Let $\mathfrak{G}$ be any Lie group with Lie algebra $\mathfrak{g}$. By Theorems 3.4.1 and 6.4.1, there is a one-one correspondence between the connected analytic subgroups of $\mathfrak{G}$ and the subalgebras of $\mathfrak{g}$. If $\mathfrak{H}_1$, $\mathfrak{H}_2$ are any two such subgroups of $\mathfrak{G}$ and $\mathfrak{h}_1$, $\mathfrak{h}_2$ the corresponding subalgebras of $\mathfrak{g}$, then it is clear that $\mathfrak{H}_1 \subseteq \mathfrak{H}_2$ if and only if $\mathfrak{h}_1 \subseteq \mathfrak{h}_2$.

Ex. If $\mathfrak{H}_1$, $\mathfrak{H}_2$ are connected analytic subgroups of a Lie group and $\mathfrak{h}_1$, $\mathfrak{h}_2$ the corresponding Lie algebras, then $\mathfrak{h}_1 \cap \mathfrak{h}_2$ is the Lie algebra of $\mathfrak{H}_1 \cap \mathfrak{H}_2$ and the subalgebra generated by $\mathfrak{h}_1$ and $\mathfrak{h}_2$ is the Lie algebra of the subgroup generated by $\mathfrak{H}_1$ and $\mathfrak{H}_2$.

6.5. Closed subgroups. Let us consider again an abstract subgroup H of a Lie group $\mathfrak{G}$. In order to define H as an analytic subgroup of $\mathfrak{G}$ we must, by Theorem 6.3.5, put a Lie group structure on H such that the identity mapping of H into $\mathfrak{G}$ is continuous. Now this may not be possible, for example, if $\mathfrak{G}$ is the group $\mathfrak{R}$ of real numbers and H is the subgroup of rational numbers, but there is one case in which H can always be defined as an analytic subgroup, and that is when H is a *closed* subgroup of $\mathfrak{G}$.

THEOREM 6.5.1. *Let $\mathfrak{G}$ be a Lie group and H a closed subgroup of $\mathfrak{G}$; then H can be defined as an analytic subgroup of $\mathfrak{G}$.*

We shall prove the theorem by constructing the Lie algebra of H. Denote the Lie algebra of $\mathfrak{G}$ by $\mathfrak{g}$, and let $\mathfrak{h}$ be the set of all $X \in \mathfrak{g}$ such that

$$\exp tX \in H \text{ for all } t \text{ in some neighbourhood of } 0.$$

We know that the mapping $t \to \exp tX$ $(X \in \mathfrak{g})$ is a homomorphism of R into $\mathfrak{G}$, and since R is generated by any neighbourhood of 0, $X \in \mathfrak{h}$ implies that $\exp tX \in H$ for all $t \in R$, so that the mapping $t \to \exp tX$ is actually a mapping of R into H in this case. To complete the proof we need the following

LEMMA. *Let Z_t be a set of elements of $\mathfrak{g}$, defined for some set of real numbers t having 0 as a limit, such that $\lim_{t \to 0} Z_t = Z$, where $Z \in \mathfrak{g}$, and suppose that $\exp tZ_t \in H$ for all t for which Z_t is defined. Then $Z \in \mathfrak{h}$.*

Proof of the lemma. If $\exp tZ_t \in H$ then

$$\exp(-tZ_t) = (\exp tZ_t)^{-1} \in H;$$

therefore we may take t to be positive. Now, given any fixed $\delta > 0$ and $t > 0$, let us define†

$$k_t = k_t(\delta) = \left[\frac{\delta}{t}\right],$$

so that k_t is a non-negative integer satisfying $k_t t \to \delta$ as $t \to 0$, for fixed δ. By hypothesis $\exp tZ_t \in H$, hence

$$\exp k_t tZ_t = (\exp tZ_t)^{k_t} \in H.$$

† $[\delta/t]$ denotes the integral part of δ/t.

As $t \to 0$, $k_t t Z_t \to \delta Z$, and so $\exp k_t t Z_t \to \exp \delta Z$. Since H is closed $\exp \delta Z \in H$ for all sufficiently small positive δ; moreover,

$$\exp(-\delta Z) = (\exp \delta Z)^{-1} \in H,$$

so that $Z \in \mathfrak{h}$ by the definition of $\mathfrak{h}$. Thus the lemma is established

To continue with the proof of Theorem 6.5.1, we show that $\mathfrak{h}$ is a subalgebra of $\mathfrak{g}$. Let $X \in \mathfrak{h}$, then $\exp tX \in H$ for all t and hence for any real α, $\exp t\alpha X \in H$ for all t, whence $\alpha X \in \mathfrak{h}$. Further, if $X, Y \in \mathfrak{h}$, then $\exp tX$, $\exp tY \in H$, and hence H contains

$$\begin{aligned} \exp tX . \exp tY &= \exp\{t(X+Y) + O(t^2)\} \\ &= \exp tZ_t, \end{aligned}$$

say, where $Z_t \to X + Y$ as $t \to 0$. Hence, by the lemma, $X + Y \in \mathfrak{h}$. Similarly, if $X, Y \in \mathfrak{h}$, then $(\exp tX, \exp tY) \in H$, and

$$\begin{aligned} (\exp tX, \exp tY) &= \exp\{t^2[X,Y] + O(t^3)\} \\ &= \exp uZ_u, \end{aligned}$$

where $u = t^2$, $Z_u \to [X, Y]$ as $u \to 0$. Again we conclude by the lemma that $[X,Y] \in \mathfrak{h}$, and this establishes $\mathfrak{h}$ as a Lie algebra.

Let $\mathfrak{H}_1$ be the connected analytic subgroup of $\mathfrak{G}$ corresponding to $\mathfrak{h}$. By Theorem 6.4.1, $\mathfrak{H}_1$ is generated by the elements

$$\exp Y \quad (Y \in \mathfrak{h}),$$

hence $\mathfrak{H}_1 \subseteq H$, and since $\mathfrak{H}_1$ is connected we can even say that $\mathfrak{H}_1 \subseteq H_0$, where H_0 is the identity component of H. We shall show that there is a nucleus U of $\mathfrak{G}$ for which $U \cap H = U \cap \mathfrak{H}_1$. Then $U \cap \mathfrak{H}_1$ is a nucleus of H_0 and it therefore generates H_0 because H_0 is connected. It follows then that $\mathfrak{H}_1 = H_0$ and this will prove that H can be defined as an analytic subgroup.

Let $X_1, \ldots, X_n$ be a basis of $\mathfrak{g}$ such that $X_1, \ldots, X_\nu$ is a basis of $\mathfrak{h}$, and use the conventions of 6.4 with regard to suffixes. If $\nu = n$, then $\mathfrak{H}_1$ is the identity component of H and there is nothing more to prove, so we may suppose $\nu < n$. Now consider any unit vector $\lambda^{\alpha'}$ on the $(n-\nu)$-dimensional unit sphere: $\Sigma(\lambda^{\alpha'})^2 = 1$. For such a vector the condition

$$\exp t\lambda^{\alpha'} X_{\alpha'} \in H \tag{19}$$

cannot hold for arbitrarily small t, since otherwise $\lambda^{\alpha'} X_{\alpha'} \in \mathfrak{h}$, which would contradict the choice of the X_i. Let $t(\lambda)$ be the greatest lower bound of positive values of t for which (19) holds, then $t(\lambda) > 0$ for each $\lambda^{\alpha'}$. Moreover, at any point $\lambda_0^{\alpha'}$ we have

$$\exp \tfrac{1}{2}t(\lambda_0)\,\lambda_0^{\alpha'} X_{\alpha'} \notin H.$$

Because H is closed it follows that in a suitably chosen neighbourhood of the point λ_0 the function $t(\lambda)$ has a positive lower bound. If we apply Lemma 6.1.2, we find that $t(\lambda)$ has a positive lower bound over the whole of the $(n-\nu)$-dimensional sphere:

$$t(\lambda) \geqslant a > 0 \text{ for all } \lambda^{\alpha'} \text{ such that } \Sigma(\lambda^{\alpha'})^2 = 1.$$

Now define a chart at e by the equations

$$(\exp x^{\alpha} X_{\alpha} . \exp x^{\alpha'} X_{\alpha'})^i = x^i, \tag{20}$$

and put $U = \{x \in \mathfrak{G} \mid \Sigma(x^i)^2 < a^2\}$. Then we assert that

$$U \cap H = U \cap \mathfrak{H}_1.$$

Since $\mathfrak{H}_1 \subseteq H$, it is enough to prove $U \cap H \subseteq U \cap \mathfrak{H}_1$. Let $x \in U \cap H$, say $x = x_1 x_2$, where $x_1 = \exp x^{\alpha} X_{\alpha}$, $x_2 = \exp x^{\alpha'} X_{\alpha'}$. By definition, $x \in U$, so that $\Sigma(x^{\alpha})^2 + \Sigma(x^{\alpha'})^2 < a^2$, whence $x_1, x_2 \in U$. Moreover, $x_1 \in \mathfrak{H}_1 \subseteq H$; this shows that $x_2 = x_1^{-1} x \in H$, i.e. $x_2 \in U \cap H$. If x_2 were different from e, then $\{\Sigma(x^{\alpha'})^2\}^{\frac{1}{2}} = b > 0$, and $b < a$, because $x_2 \in U$. Write $\lambda^{\alpha'} = x^{\alpha'}/b$, then $\Sigma(\lambda^{\alpha'})^2 = 1$ and

$$\exp b\lambda^{\alpha'} X_{\alpha'} = x_2 \in H.$$

But this contradicts the fact that $b < a$. Hence we must have $x_2 = e$ and $x = x_1 \in U \cap \mathfrak{H}_1$. Thus $U \cap H \subseteq U \cap \mathfrak{H}_1$, as we wished to show; the proof is now complete.

In the course of the proof we have shown that if $\mathfrak{G}$ is any Lie group and $\mathfrak{H}$ a closed subgroup of $\mathfrak{G}$—which may be taken to be analytic, by the theorem just proved—then we can define a chart at e in $\mathfrak{G}$ by the equations (20) such that in a suitable nucleus U of $\mathfrak{G}$, $x \in U \cap \mathfrak{H}$ implies $x^{\alpha'} = 0$.

We are going to consider the quotient space $\mathfrak{G}/\mathfrak{H}$ and to show that it can be defined as an analytic manifold in a natural way. Let ϕ be the natural mapping of $\mathfrak{G}$ onto $\mathfrak{G}/\mathfrak{H}$. Now choose a nucleus V of $\mathfrak{G}$ such that $VV^{-1} \subseteq U$ and $V^{-1}V \subseteq U$, then on the subset of V defined by

$$x^{\alpha} = 0 \quad (x \in V) \tag{21}$$

the mapping ϕ is one-one. For if $x, y \in V$, $x^\alpha = y^\alpha = 0$, and $x^\phi = y^\phi$ then $x = hy$, where $h \in \mathfrak{H}$; we have $h = xy^{-1} \in VV^{-1} \subseteq U$, whence $h = \exp h^\alpha X_\alpha$. Thus

$$\exp x^{\alpha'} X_{\alpha'} = \exp h^\alpha X_\alpha . \exp y^{\alpha'} X_{\alpha'},$$

and by the uniqueness of the coordinates, $h = e$ and $x = y$. Let ψ be the inverse mapping from V^ϕ, $= W$ say, to the region defined by (21), and introduce coordinates on W by assigning to $w \in W$ the $n - \nu$ last coordinates of $w^\psi (\in V)$.

If we regard $\mathfrak{G}/\mathfrak{H}$ as a topological space (the topology being that induced by $\mathfrak{G}$) then ϕ is an open continuous mapping; hence its restriction to W^ψ—where it is one-one—is a homeomorphism, in other words, ψ is also continuous. With the chart on W defined by ψ as described above, ϕ and ψ are even analytic. Now the right multiplications by elements of $\mathfrak{G}$ define homeomorphisms of $\mathfrak{G}/\mathfrak{H}$ and in this way we get a chart at each point of $\mathfrak{G}/\mathfrak{H}$. If two charts overlap, say $u \in Wa \cap Wb$ $(a, b \in \mathfrak{G})$, let

$$u^\psi = xa = yb \quad (x, y \in V),$$

then

$$ab^{-1} = x^{-1}y \in V^{-1}V \subseteq U,$$

and

$$(u)_a^{\alpha'} = x^{\alpha'} = (y . ba^{-1})^{\alpha'},$$

$$(u)_b^{\alpha'} = y^{\alpha'} = (x . ab^{-1})^{\alpha'},$$

which shows that the two charts are analytically related at u. Now u was any point of $\mathfrak{G}/\mathfrak{H}$ common to two charts, so we have found an analytic family of charts covering $\mathfrak{G}/\mathfrak{H}$, which can be used to define $\mathfrak{G}/\mathfrak{H}$ as an analytic manifold. The analytic structure so defined is unique, since it is determined near the coset $\mathfrak{H}$ by the fact that ϕ and ψ are analytic, and elsewhere because the right multiplications are analytic homeomorphisms.

These facts may be summed up in

THEOREM 6.5.2. *Let $\mathfrak{G}$ be a Lie group, $\mathfrak{H}$ a closed analytic subgroup and denote by ϕ the natural mapping of $\mathfrak{G}$ onto $\mathfrak{G}/\mathfrak{H}$. Then $\mathfrak{G}/\mathfrak{H}$ can be defined as an analytic manifold in such a way that*

(i) *the mapping ϕ is analytic,*

(ii) *there is a mapping ψ of a neighbourhood W of the coset $\mathfrak{H}$ of $\mathfrak{G}/\mathfrak{H}$ into $\mathfrak{G}$ which is analytic and such that $\psi\phi$ is the identity mapping on W.*

The analytic structure on $\mathfrak{G}/\mathfrak{H}$ is uniquely determined by (i) *and* (ii), *and the manifold topology of $\mathfrak{G}/\mathfrak{H}$ is that induced by the topology of $\mathfrak{G}$.*

The mapping ψ is called a *local cross-section* of the mapping ϕ; thus Theorem 6.5.2 asserts *inter alia* the existence of a local cross-section under the conditions stated.

Ex. 1. If $\mathfrak{H}$ is an analytic subgroup of a Lie group $\mathfrak{G}$, and $\mathfrak{H}$ is a closed subset of $\mathfrak{G}$, then the topology of $\mathfrak{H}$ is that induced by the topology of $\mathfrak{G}$.

Ex. 2. Let $\mathfrak{T}^2$ be the 2-dimensional torus group, with coordinates x, y. Then for irrational α, the set of points defined by $x=t,\ y=\alpha t \pmod 1$ forms an analytic subgroup of $\mathfrak{T}^2$ which is not closed. Moreover, its topology is different from the topology induced by $\mathfrak{T}^2$.

6.6. Homomorphisms and quotient groups. Next we consider homomorphisms of Lie groups and quotients of Lie groups and show that they have their counterparts in the Lie algebra of the group. We must begin by defining the analogous concepts for a Lie algebra:

A mapping $X \to X'$ of a Lie algebra $\mathfrak{g}$ into a Lie algebra $\mathfrak{h}$ is called a *homomorphism*, if it is linear:

$$(\alpha X + \beta Y)' = \alpha X' + \beta Y' \quad (\alpha, \beta \in R), \tag{22}$$

and, moreover,

$$[X, Y]' = [X', Y'] \quad (X, Y \in \mathfrak{g}). \tag{23}$$

Thus an isomorphism of $\mathfrak{g}$ onto $\mathfrak{h}$, as defined in 3.1, is simply a homomorphism which has an inverse. A homomorphism of $\mathfrak{g}$ into $\mathfrak{h}$ is an isomorphism of $\mathfrak{g}$ with a subalgebra of $\mathfrak{h}$, if and only if 0 is the only element of $\mathfrak{g}$ which maps into the 0 of $\mathfrak{h}$. In the case of the general homomorphism the elements of $\mathfrak{g}$ which map into 0 form a set $\mathfrak{n}$ which is called the *kernel* of the homomorphism. By (22), $X, Y \in \mathfrak{n}$ implies $\alpha X + \beta Y \in \mathfrak{n}$, so that $\mathfrak{n}$ is a linear space, and by (23), if $Z \in \mathfrak{n}$, then for any $X \in \mathfrak{g}$,

$$[X, Z]' = [X', Z'] = [X', 0] = 0,$$

so that

$$[X, Z] \in \mathfrak{n} \quad \text{for all } X \in \mathfrak{g},\ Z \in \mathfrak{n}. \tag{24}$$

A subspace of $\mathfrak{g}$ satisfying (24) is called an *ideal*† of $\mathfrak{g}$; thus the kernel of any homomorphism of $\mathfrak{g}$ is an ideal of $\mathfrak{g}$.

Conversely, let $\mathfrak{k}$ be any ideal of $\mathfrak{g}$; then we can form the factor space $\mathfrak{g}/\mathfrak{k}$, which consists of the residue classes mod $\mathfrak{k}$, and on $\mathfrak{g}/\mathfrak{k}$ define a multiplication as follows:

If $X^*, Y^* \in \mathfrak{g}/\mathfrak{k}$, let X, Y be representatives of X^*, Y^* respectively, and denote by $[X, Y]^*$ the coset containing $[X, Y]$. This coset depends only on X^*, Y^* and not on X, Y, for if we replace X, Y by other representatives $X+Z_1$, $Y+Z_2$, where $Z_1, Z_2 \in \mathfrak{k}$, then

$$[X+Z_1,\ Y+Z_2]=[X,Y]+[X,Z_2]-[Y+Z_2,Z_1],$$

whence $[X+Z_1,\ Y+Z_2]^*=[X,Y]^*$. We may therefore define

$$[X^*,Y^*]=[X,Y]^*; \tag{25}$$

it is easily verified that with this multiplication $\mathfrak{g}/\mathfrak{k}$ becomes a Lie algebra. Thus for any ideal $\mathfrak{k}$ in $\mathfrak{g}$, the mapping $X \rightarrow X^*$, where X^* denotes the residue class mod $\mathfrak{k}$ of X, is a mapping of $\mathfrak{g}$ onto the Lie algebra $\mathfrak{g}/\mathfrak{k}$ just defined. Equation (25) and the equation

$$(\alpha X+\beta Y)^*=\alpha X^*+\beta Y^*$$

show that this mapping is a homomorphism; its kernel is the coset containing 0, i.e. the ideal $\mathfrak{k}$. We call $\mathfrak{g}/\mathfrak{k}$ the *quotient algebra* of $\mathfrak{g}$ by $\mathfrak{k}$ and $X \rightarrow X^*$ the *natural homomorphism* of $\mathfrak{g}$ onto $\mathfrak{g}/\mathfrak{k}$; then these results may be stated as

THEOREM 6.6.1. *If* $\mathfrak{g}$ *is any Lie algebra and* $\mathfrak{k}$ *an ideal in* $\mathfrak{g}$, *then* $\mathfrak{g}/\mathfrak{k}$ *can be defined as a Lie algebra*; $\mathfrak{g}/\mathfrak{k}$ *is a homomorphic image of* $\mathfrak{g}$ *under the natural homomorphism, and the kernel of this homomorphism is* $\mathfrak{k}$.

The corresponding result for abstract groups is of course well known (cf. also 2.3). The result holds in fact for any linear algebra, or more generally, for any group with operators.

Let $\mathfrak{G}$ and $\mathfrak{H}$ be any Lie groups with Lie algebras $\mathfrak{g}$ and $\mathfrak{h}$ respectively, and let Φ be an analytic homomorphism of $\mathfrak{G}$

† We note that if (24) holds then we have also $[Z, X] \in \mathfrak{n}$ for all $X \in \mathfrak{g}$, $Z \in \mathfrak{n}$, since $[Z, X] = -[X, Z]$. In general linear algebras, where the products ab and ba are not, in general, linearly dependent, this distinction leads to left and right ideals.

into $\mathfrak{H}$. We may use Φ to define a mapping of $\mathfrak{g}$ into $\mathfrak{h}$ as follows: Let $X \in \mathfrak{g}$, $X \neq 0$, then $\exp tX$ is a 1-parameter subgroup of $\mathfrak{G}$ and hence $(\exp tX)^{\Phi}$ is a 1-parameter subgroup of $\mathfrak{H}$, or reduces to the unit element of $\mathfrak{H}$. In either case we can write

$$(\exp tX)^{\Phi} = \exp t\bar{X} \quad (\bar{X} \in \mathfrak{h}).$$

This formula even holds, with $\bar{X} = 0$, when $X = 0$, and it defines $\bar{X}$ uniquely in terms of X. In order to obtain the coordinates of $\bar{X}$, let us take any charts $(x^1, \ldots, x^m)$, $(y^1, \ldots, y^n)$ at the unit elements e and e' of $\mathfrak{G}$ and $\mathfrak{H}$ respectively. For convenience we suppose that in each case the unit element corresponds to the origin. The mapping Φ is then described by equations of the form

$$y^k = \Phi^k(x) \quad (\Phi^k(0) = 0).$$

Since X and $\bar{X}$ are left-invariant it is enough to express $\bar{X}$ in terms of X at a single point, say at e. Let

$$X_e = \lambda^i \frac{\partial}{\partial x^i}, \quad \bar{X}_{e'} = \mu^k \frac{\partial}{\partial y^k},$$

be the values of X and $\bar{X}$ at e and e', where the repeated suffixes imply summations over the appropriate ranges. Near e' we have

$$(\exp t\bar{X})^k = t\mu^k + O(t^2),$$

while

$$\begin{aligned}((\exp tX)^{\Phi})^k &= \Phi^k(t\lambda + O(t^2)) \\ &= \left[\frac{\partial \Phi^k}{\partial x^i}\right]_{x=e} t\lambda^i + O(t^2).\end{aligned}$$

A comparison shows that

$$\mu^k = \lambda^i \left[\frac{\partial \Phi^k}{\partial x^i}\right]_{x=e}.$$

In the case where Φ has an analytic inverse, this equation just represents the coordinates of the infinitesimal transformation $X^{d\Phi}$. We shall therefore denote $\bar{X}$ by $X^{d\Phi}$; thus the definition of $X^{d\Phi}$ reads

$$\exp tX^{d\Phi} = (\exp tX)^{\Phi} \quad (X \in \mathfrak{g}). \tag{26}$$

We note that this definition applies only when Φ is a homomorphism. It is precisely this fact which enables us to assert that $X^{d\Phi}$ as defined by (26) is in $\mathfrak{h}$, and is therefore determined by its value at e'. This in turn enables us to dispense with the fina mapping $\{\Phi^{-1}\}$ in the original definition of $d\Phi$ (1.8) and sc extend the definition to mappings Φ which are not necessarily invertible. Since in any case Φ must be a homomorphism, it is enough to assume Φ to be continuous, because it is then analytic, by Theorem 6.3.2. As in 1.8, we can prove

THEOREM 6.6.2. *If I is the identity mapping of a Lie group $\mathfrak{G}$ onto itself, then I is a continuous homomorphism, and dI is the identity mapping of $\Lambda(\mathfrak{G})$ onto itself. If Φ and Ψ are continuous homomorphisms of $\mathfrak{G}$ into $\mathfrak{H}$ and $\mathfrak{H}$ into $\mathfrak{K}$ respectively, where $\mathfrak{G}, \mathfrak{H}$ and $\mathfrak{K}$ are any Lie groups, then $\Phi . \Psi$ is a continuous homomorphism of $\mathfrak{G}$ into $\mathfrak{K}$, and*

$$d(\Phi . \Psi) = d\Phi . d\Psi.$$

In view of the correspondence between Lie groups and Lie algebras we are led to expect that the mapping $d\Phi$ defined by the homomorphism Φ is a homomorphism between the corresponding Lie algebras, and this is in fact the case.

THEOREM 6.6.3. *If Φ is a continuous homomorphism of a Lie group $\mathfrak{G}$ into a Lie group $\mathfrak{H}$ then the mapping $d\Phi$ is a homomorphism of $\Lambda(\mathfrak{G})$ into $\Lambda(\mathfrak{H})$. If the kernels of Φ and $d\Phi$ are denoted by $\mathfrak{N}$ and $\mathfrak{n}$ respectively, then $\mathfrak{N}$ is a closed normal subgroup of $\mathfrak{G}$, $\mathfrak{n}$ an ideal of $\Lambda(\mathfrak{G})$ and $\mathfrak{n}$ is the Lie algebra of $\mathfrak{N}$. The image $\mathfrak{G}_1$ of $\mathfrak{G}$ under Φ is an analytic subgroup of $\mathfrak{H}$ and its Lie algebra is $\mathfrak{g}_1$, the image of $\Lambda(\mathfrak{G})$ under $d\Phi$. In particular, if Φ maps $\mathfrak{G}$ onto $\mathfrak{H}$, $d\Phi$ maps $\Lambda(\mathfrak{G})$ onto $\Lambda(\mathfrak{H})$.*

In the proof we denote $X^{d\Phi}$ by $\bar{X}$ for short, and put $\mathfrak{g} = \Lambda(\mathfrak{G})$, $\mathfrak{h} = \Lambda(\mathfrak{H})$. Then if $X, Y \in \mathfrak{g}$,

$$\begin{aligned}
&\exp t(\overline{X+Y} - \bar{X} - \bar{Y}) \\
&\quad = \exp t(\overline{X+Y}) . \exp(-t\bar{X}) . \exp(-t\bar{Y}) . \exp O(t^2) \\
&\quad = \{\exp t(X+Y)\}^{\Phi} \exp(-tX)^{\Phi} \exp(-tY)^{\Phi} \exp O(t^2) \\
&\quad = \{\exp t(X+Y) . \exp(-tX) . \exp(-tY)\}^{\Phi} \exp O(t^2) \\
&\quad = \exp O(t^2)^{\Phi} \exp O(t^2) \\
&\quad = \exp O(t^2).
\end{aligned}$$

Here we have used Theorem 6.1.4, (i) and (iv), as well as the fact that Φ is analytic (Theorem 6.3.2). Taking logarithms, we find

$$t(\overline{X+Y}-\bar{X}-\bar{Y})=O(t^2).$$

Hence, when we divide by t and let $t\to 0$, $\overline{X+Y}=\bar{X}+\bar{Y}$, i.e.

$$(X+Y)^{d\Phi}=X^{d\Phi}+Y^{d\Phi}.$$

Clearly $(\alpha X)^{d\Phi}=\alpha X^{d\Phi}$, and finally, by Theorem 6.1.4, (iii), (iv), (v),

$$\begin{aligned}
\exp\,(t^2\overline{[X,Y]}-t^2[\bar{X},\bar{Y}])
&=\exp t^2\overline{[X,Y]}\,.\,(\exp t\bar{X},\,\exp t\bar{Y})^{-1}\,.\exp O(t^3)\\
&=\{\exp t^2[X,Y]\}^{\Phi}\,((\exp tX)^{\Phi},(\exp tY)^{\Phi})^{-1}\,.\exp O(t^3)\\
&=\{\exp t^2[X,Y]\,.\,(\exp tX,\,\exp tY)^{-1}\}^{\Phi}\exp O(t^3)\\
&=\exp O(t^3)^{\Phi}\exp O(t^3)\\
&=\exp O(t^3).
\end{aligned}$$

Hence $t^2(\overline{[X,Y]}-[\bar{X},\bar{Y}])=O(t^3)$, and as before we deduce that

$$[X,Y]^{d\Phi}=[X^{d\Phi},Y^{d\Phi}].$$

This proves that $d\Phi$ is a homomorphism of $\mathfrak{g}$ into $\mathfrak{h}$. Now let $\mathfrak{N}$ be the kernel of Φ and $\mathfrak{n}$ the kernel of $d\Phi$. Then $\mathfrak{N}$ is a normal subgroup of $\mathfrak{G}$, and $\mathfrak{N}$ is closed as the kernel of a continuous homomorphism; similarly $\mathfrak{n}$ is an ideal of $\mathfrak{g}$. Let $\mathfrak{n}_1$ be the subalgebra of $\mathfrak{g}$ belonging to the subgroup $\mathfrak{N}$ and write e' for the unit-element of $\mathfrak{H}$. An element Z of $\mathfrak{g}$ lies in $\mathfrak{n}_1$ if and only if $(\exp tZ)^{\Phi}=e'$ for all t, i.e. $\exp tZ^{d\Phi}=e'$. But this is true precisely when $Z^{d\Phi}=0$, which is the condition for Z to lie in $\mathfrak{n}$; thus $\mathfrak{n}_1=\mathfrak{n}$.

If $\mathfrak{G}_1$, $\mathfrak{g}_1$ are the images of $\mathfrak{G}$, $\mathfrak{g}$ under the mappings Φ, $d\Phi$ respectively, then it is easily seen that $\mathfrak{g}_1$ is a subalgebra of $\mathfrak{h}$, and the connected analytic subgroup of $\mathfrak{H}$ which is associated with $\mathfrak{g}_1$ is generated by the elements $\exp X^{d\Phi}=(\exp X)^{\Phi}$ (Theorem 6.5.1), and is therefore contained in $\mathfrak{G}_1$. On the other hand, this subgroup contains a nucleus of $\mathfrak{G}_1$ and therefore coincides with the identity component of $\mathfrak{G}_1$ (Theorem 2.8.3). This shows that $\mathfrak{g}_1$ is the Lie algebra of $\mathfrak{G}_1$ and, incidentally, that $\mathfrak{G}_1$ is an analytic subgroup of $\mathfrak{G}$, a fact which is not hard to see directly. Thus all the assertions made in Theorem 6.6.3 are proved.

We note the special case of Theorem 6.6.3, where $\mathfrak{H} = \mathfrak{G}$. A homomorphism of a group (or an algebra) into itself is called an *endomorphism*, so to every continuous endomorphism Φ of $\mathfrak{G}$ corresponds an endomorphism $d\Phi$ of $\Lambda(\mathfrak{G})$. This applies in particular to the *automorphisms* of $\mathfrak{G}$, which are just the invertible endomorphisms of $\mathfrak{G}$, i.e. the endomorphisms of $\mathfrak{G}$ which are also isomorphisms of $\mathfrak{G}$ with itself.

THEOREM 6.6.4. *Let $\mathfrak{G}$ be a Lie group and $\mathfrak{N}$ a closed normal subgroup, and let their respective Lie algebras be $\mathfrak{g}$ and $\mathfrak{n}$, where $\mathfrak{n} \subseteq \mathfrak{g}$. Then the quotient group $\mathfrak{G}/\mathfrak{N}$ with its analytic structure is a Lie group whose Lie algebra is isomorphic to $\mathfrak{g}/\mathfrak{n}$.*†

Proof. We know from Theorem 6.5.2 that $\mathfrak{G}/\mathfrak{N}$ is an analytic manifold and that the natural homomorphism ϕ of $\mathfrak{G}$ onto $\mathfrak{G}/\mathfrak{N}$ is analytic. Let ψ be a local cross-section of ϕ; if x, y, z lie in a—suitably chosen—nucleus of $\mathfrak{G}/\mathfrak{N}$ and $xy = z$, then

$$z = x^{\psi\phi} y^{\psi\phi} = (x^{\psi} y^{\psi})^{\phi},$$

which is analytic in x and y. Therefore $\mathfrak{G}/\mathfrak{N}$ is a Lie group with respect to the analytic structure considered.

The kernel of the mapping ϕ is $\mathfrak{N}$, and by Theorem 6.6.3, the kernel of $d\phi$ is the Lie algebra $\mathfrak{n}$ of $\mathfrak{N}$; moreover, ϕ is a continuous homomorphism of $\mathfrak{G}$ onto $\mathfrak{G}/\mathfrak{N}$ and so $d\phi$ is a homomorphism of $\mathfrak{g}$ onto the Lie algebra $\mathfrak{h}$, say, of $\mathfrak{G}/\mathfrak{N}$. Since $\mathfrak{n}$ is the kernel of $d\phi$, $\mathfrak{g}/\mathfrak{n}$ is isomorphic to $\mathfrak{h}$, and this completes the proof.

6.7. The general linear group as a Lie group.

An important example of a Lie group is the general linear group $\mathfrak{GL}(n, R)$. We have already seen in 1.2 that this is an analytic manifold, and from 2.6 we know that it is a Lie group. We shall denote this group by $\mathfrak{GL}$ for short. When the elements are expressed as non-singular $n \times n$ matrices (x_{ij}), the multiplication law takes the form

$$(xy)_{ik} = x_{ij} y_{jk}.‡$$

† This implies that $\mathfrak{n}$ is an ideal of $\mathfrak{g}$. The converse, that $\mathfrak{N}$ is a normal subgroup of $\mathfrak{G}$ whenever its Lie algebra $\mathfrak{n}$ is an ideal of $\mathfrak{g}$, is less obvious. Both facts will be proved together in Theorem 6.8.2.

‡ In this section we use pairs of suffixes, each running from 1 to n, since the dimension of $\mathfrak{GL}$ is n^2. We write all suffixes below, but keep the summation convention.

To obtain the transformation functions of $\mathfrak{GL}$, we have

$$\left[\frac{\partial (xy)_{ij}}{\partial y_{rs}}\right]_{y=e} = \left[\frac{\partial}{\partial y_{rs}}(x_{ik}\,y_{kj})\right]_{y=e} = \delta_{js}x_{ir},$$

where δ_{js} is the Kronecker delta. Hence a basis for the Lie algebra $\mathfrak{gl}$, say, of $\mathfrak{GL}$ is given by

$$X_{rs} = \delta_{js}x_{ir}\frac{\partial}{\partial x_{ij}} = x_{ir}\frac{\partial}{\partial x_{is}}. \tag{27}$$

Let us calculate the multiplication table for this Lie algebra. We have

$$\begin{aligned} X_{rs}X_{uv} &= x_{ir}\frac{\partial}{\partial x_{is}}\left(x_{ju}\frac{\partial}{\partial x_{jv}}\right) \\ &= x_{ir}\delta_{ij}\delta_{su}\frac{\partial}{\partial x_{jv}} + \ldots \\ &= x_{ir}\delta_{su}\frac{\partial}{\partial x_{iv}} + \ldots \\ &= \delta_{su}X_{rv} + \ldots, \end{aligned}$$

where the dots indicate second-order terms, which we know will cancel. Thus

$$[X_{rs}, X_{uv}] = \delta_{us}X_{rv} - \delta_{rv}X_{us}. \tag{28}$$

Now let M_n be the linear associative algebra of $n \times n$ matrices over R. The algebra M_n has a basis E_{ij} $(i, j = 1, \ldots, n)$, where E_{ij} is the matrix which has 1 in the (i, j)th place and zeros everywhere else. The multiplication table of M_n relative to this basis is

$$E_{rs}E_{uv} = \delta_{us}E_{rv}. \tag{29}$$

The set M_n may also be regarded as a Lie algebra, if we define it as a linear space in the usual way, but with the multiplication

$$[A, B] = AB - BA \quad (A, B \in M_n). \tag{30}$$

(cf. 3.1, Ex. 1). The E_{ij} again form a basis for this Lie algebra—since the underlying vector space is the same as for the associative algebra—and the multiplication table is, by (29) and (30),

$$[E_{rs}, E_{uv}] = \delta_{us}E_{rv} - \delta_{rv}E_{us};$$

a comparison with (28) shows that this algebra is isomorphic with $\mathfrak{gl}$. Thus we obtain

THEOREM 6.7.1. *The space of* $n \times n$ *matrices over* R *with the multiplication* (30) *forms a Lie algebra isomorphic to the Lie algebra* $\mathfrak{gl}$ *of the group* $\mathfrak{GL}(n, R)$.

This theorem is often expressed by saying that the Lie algebra of the group $\mathfrak{GL}(n, R)$ is the Lie algebra of $n \times n$ matrices over R. For that reason we shall denote this algebra—even when it is not considered in connexion with $\mathfrak{GL}$—by $\mathfrak{gl}$ or $\mathfrak{gl}(n, R)$.

Using Theorem 6.4.1, we obtain the following important consequence of Theorem 6.7.1:

THEOREM 6.7.2. *If* L *is any Lie algebra of* $n \times n$ *matrices (relative to the multiplication* (30)), *then there is a Lie group whose algebra is isomorphic to* L.

This theorem goes some way towards providing a global converse to Lie's Third Theorem. In fact this converse may be proved by using Theorem 6.7.2 together with Ado's theorem, which states that any finite dimensional Lie algebra over R is isomorphic to a subalgebra of $\mathfrak{gl}(n, R)$, for some n.

Finally, we consider what becomes of the exponential mapping in $\mathfrak{GL}$. In terms of the basis X_{ij} of $\mathfrak{gl}(n, R)$, defined in (27) above, the general element of $\mathfrak{gl}$ is

$$\Lambda = \lambda_{ij} X_{ij}.$$

Let L denote the matrix (λ_{ij}), then we can form the series

$$\sum_{k=0}^{\infty} \frac{L^k}{k!}. \tag{31}$$

This series is convergent in the sense that the (i, j)th element of the partial sum of k terms tends to a limit as $k \to \infty$. It can also be verified that

$$\sum_h \frac{L^h}{h!} \sum_k \frac{(-L)^k}{k!} = I,$$

where I denotes the unit matrix. This shows that the matrix defined by the series (31) is non-singular and therefore belongs to $\mathfrak{GL}$. On the other hand, we can form the matrix $\exp \Lambda$ by applying the exponential mapping to Λ. We shall show that

$$\exp \Lambda = \sum_k \frac{L^k}{k!}. \tag{32}$$

For we have, by definition, $\Lambda = \lambda_{rs} X_{rs} = \lambda_{rs} x_{tr} \dfrac{\partial}{\partial x_{ts}}$. Hence

$$\begin{aligned}\Lambda^k x_{uv} &= \lambda_{r_1 s_1} \dots \lambda_{r_k s_k} x_{t_1 r_1} \frac{\partial}{\partial x_{t_1 s_1}} \dots x_{t_k r_k} \frac{\partial}{\partial x_{t_k s_k}} x_{uv} \\ &= \lambda_{r_1 s_1} \dots \lambda_{r_k s_k} \delta_{u t_k} \delta_{v s_k} \delta_{t_k t_{k-1}} \delta_{r_k s_{k-1}} \dots \delta_{t_2 t_1} \delta_{r_2 s_1} x_{t_1 r_1} \\ &= \lambda_{r_1 r_2} \lambda_{r_2 r_3} \dots \lambda_{r_k v} x_{u r_1} \\ &= x_{u r_1} (L^k)_{r_1 v},\end{aligned}$$

where $(L^k)_{ij}$ denotes the (i,j)th element of the matrix L^k. Now put $x = I$, the unit element of $\mathfrak{GL}$, then

$$\begin{aligned}[\Lambda^k x_{uv}]_{x=I} &= \delta_{ur} (L^k)_{rv} \\ &= (L^k)_{uv}.\end{aligned}$$

Hence $(\exp \Lambda)_{uv} = \left(\sum_k \dfrac{\Lambda^k}{k!} x_{uv} \right)_{x=I} = \left(\sum_k \dfrac{L^k}{k!} \right)_{uv}$, which proves (32).

6.8. The adjoint representation of a Lie group. Let $\mathfrak{G}$ again be an arbitrary Lie group of dimension n. By a *representation* of *degree* r of $\mathfrak{G}$ we shall understand a continuous homomorphism of $\mathfrak{G}$ into $\mathfrak{GL}(r, R)$:

$$x \to \rho(x).$$

Since the homomorphism is continuous, it is analytic, by Theorem 6.3.2. We shall not take up general representations of Lie groups, but briefly consider the *adjoint representation*, a representation which exists for every Lie group and which is of degree equal to the dimension of the group.

Let $\mathfrak{G}$ be a Lie group of dimension n, and $\mathfrak{g}$ its Lie algebra. Every continuous endomorphism ϕ of $\mathfrak{G}$ defines an endomorphism $d\phi$ of $\mathfrak{g}$: $\quad (\exp tX)^\phi = \exp tX^{d\phi}$.

If ψ is a second continuous endomorphism of $\mathfrak{G}$, then

$$d\phi . d\psi = d(\phi . \psi)$$

and the identity mapping I on $\mathfrak{G}$ corresponds to the identity mapping on $\mathfrak{g}$: $dI = I$, by Theorem 6.6.2. In particular, if α is a topological automorphism of $\mathfrak{G}$, then

$$d\alpha . d\alpha^{-1} = d(\alpha . \alpha^{-1}) = I \quad \text{and} \quad d\alpha^{-1} . d\alpha = I,$$

from which it follows that $d\alpha$ is an automorphism of $\mathfrak{g}$.

Now each inner automorphism of $\mathfrak{G}$:

$$\alpha_x: a \to x^{-1}ax \quad (a \in \mathfrak{G})$$

is clearly continuous and therefore defines an automorphism $d\alpha$ of $\mathfrak{g}$. In terms of a basis $X_i (i = 1, \ldots, n)$ of $\mathfrak{g}$ we have

$$X_i^{d\alpha_x} = \rho_i^j(x) X_j, \tag{33}$$

where the $\rho_i^j(x)$ are real numbers depending on x. We may define a matrix $\rho(x)$ whose (ij)th element is $\rho_i^j(x)$, and with this definition we have

$$\rho(x)\rho(y) = \rho(xy), \quad \rho(e) = I. \tag{34}$$

For by (33),
$$\begin{aligned} \rho_i^k(xy) X_k = X_i^{d\alpha_{xy}} &= X_i^{d\alpha_x d\alpha_y} \\ &= (\rho_i^j(x) X_j)^{d\alpha_y} \\ &= \rho_i^j(x) \rho_j^k(y) X_k, \end{aligned}$$

and $X_i^{d\alpha_e} = X_i$, which proves the equations (34). To show that we have a representation we still have to verify that the mapping $x \to \rho(x)$ is continuous. From the definition of α_x we have

$$\exp A^{d\alpha_x} = (\exp A)^{\alpha_x} = x^{-1}(\exp A)\, x \quad (A \in \mathfrak{g}),$$

hence $\exp A^{d\alpha_x}$, and with it $A^{d\alpha_x}$, is a continuous function of x; this proves that $\rho(x)$ depends continuously on x.

The representation
$$x \to \rho(x)$$

of $\mathfrak{G}$ defined in this way is called the *adjoint representation* of $\mathfrak{G}$. Thus the matrices of the adjoint representation operate on the space formed by $\mathfrak{g}$. It follows that the degree of the adjoint representation is n.

Now we observe that the adjoint representation ρ of $\mathfrak{G}$ is a continuous homomorphism of $\mathfrak{G}$ into $\mathfrak{GL}(n, R)$; therefore there corresponds to ρ a homomorphism $d\rho$ of $\mathfrak{g}$ into $\mathfrak{gl}(n, R)$. Thus with each $X \in \mathfrak{g}$, an element $X^{d\rho}$ of $\mathfrak{gl}(n, R)$ is associated, i.e. an $n \times n$ matrix, which we may interpret as a linear transformation of an n-dimensional vector space. It is natural to take $\mathfrak{g}$ itself to be this space, and when this is done the mapping $X^{d\rho}$ can be described as follows:

THEOREM 6.8.1. *If ρ is the adjoint representation of the Lie group $\mathfrak{G}$, then for each $X \in \mathfrak{g} = \Lambda(\mathfrak{G})$, the mapping $d\rho$ defines the linear mapping*
$$X^{d\rho}: A \to [A, X] \quad (A \in \mathfrak{g})$$
of $\mathfrak{g}$ into itself.

In the proof we shall regard $\rho(x)$, not as a matrix, but as a near mapping of $\mathfrak{g}$ into itself; as such we shall write it on the ight. This does not affect the proof and it simplifies the notation. 'hus, if $x = \exp tX \in \mathfrak{G}$, and if A is any element of $\mathfrak{g}$, then

$$A^{d\alpha_x} = A\rho(x),$$

nd hence,

$$\begin{aligned} \exp tA^{d\alpha_x} &= \exp tA\,\rho(x) \\ &= \exp tA\,\rho\,(\exp tX) \\ &= \exp\{tA(1+tX^{d\rho}) + O(t^3)\}. \end{aligned} \tag{35}$$

On the other hand, by the definition of $d\alpha_x$,

$$\begin{aligned} \exp tA^{d\alpha_x} &= (\exp tA)^{\alpha_x} \\ &= x^{-1}(\exp tA)\,x \\ &= \exp tA\,.\,(\exp tA, \exp tX) \\ &= \exp tA\,.\exp\{t^2[A,X] + O(t^3)\} \\ &= \exp\{tA + t^2[A,X] + O(t^3)\}. \end{aligned} \tag{36}$$

Here we have used Theorem 3.7.2 (iii) and (ii) to transform the exponentials. A comparison of (35) and (36) yields

$$tA + t^2A\,.\,X^{d\rho} + O(t^3) = tA + t^2[A,X] + O(t^3).$$

When we subtract tA, divide by t^2 and let $t \to 0$, we obtain

$$A\,.\,X^{d\rho} = [A,X],$$

which establishes the theorem.

Thus the mapping $X \to X^{d\rho}$ is a homomorphism of $\mathfrak{g}$ into $\mathfrak{gl}(n,R)$, which is called the *adjoint representation* of $\mathfrak{g}$.† It is easy to verify directly that it is a homomorphism: By definition,

$$A\,.\,X^{d\rho}\,.\,Y^{d\rho} = [[A,X],Y],$$

hence

$$\begin{aligned} A\,.\,[X^{d\rho}, Y^{d\rho}] &= [[A,X],Y] - [[A,Y],X] \\ &= [A,[X,Y]] \quad \text{by the Jacobi identity} \\ &= A\,.\,[X,Y]^{d\rho}, \end{aligned}$$

Thus $[X,Y]^{d\rho} = [X^{d\rho}, Y^{d\rho}]$, as we wished to show.

† More generally, any homomorphism of a Lie algebra L over R into a Lie algebra $\mathfrak{gl}(r,R)$ for some r is called a *representation* of L.

As an application we prove

THEOREM 6.8.2. *Let $\mathfrak{G}$ be a connected Lie group with Lie algebr[a] $\mathfrak{g}$, and $\mathfrak{H}$ a connected analytic subgroup of $\mathfrak{G}$ with Lie algebra contained in $\mathfrak{g}$. Then $\mathfrak{h}$ is an ideal in $\mathfrak{g}$ if and only if $\mathfrak{H}$ is norma[l] in $\mathfrak{G}$.*

Proof. If $\mathfrak{H}$ is normal in $\mathfrak{G}$, then for all $X \in \mathfrak{g}$, $A \in \mathfrak{h}$ and all real t

$$tA.\rho(\exp tX) \in \mathfrak{h}.$$

As in the proof of Theorem 6.8.1, we have

$$tA.\rho(\exp tX) = tA + t^2[A,X] + O(t^3);$$

it follows that for each $t \neq 0$ there is an element $[A,X] + O(t)$ in $\mathfrak{h}$; letting $t \to 0$ we find that $[A,X] \in \mathfrak{h}$, because $\mathfrak{h}$ is a subspace of $\mathfrak{g}$. This proves that $\mathfrak{h}$ is an ideal in $\mathfrak{g}$. Conversely, if $\mathfrak{h}$ is an ideal in $\mathfrak{g}$, then $[A,X] = A.X^{d\rho} \in \mathfrak{h}$ for all $X \in \mathfrak{g}$, $A \in \mathfrak{h}$. Hence

$$\exp\{A \exp X^{d\rho}\} \in \mathfrak{H},$$

i.e. $\exp\{A\rho(x)\} \in \mathfrak{H}$, where $x = \exp X$. Now

$$\exp\{A\rho(x)\} = x^{-1}(\exp A)\,x,$$

and as X runs through $\mathfrak{g}$, $x = \exp X$ runs through a nucleus of $\mathfrak{G}$. Since $\mathfrak{G}$ is connected, this nucleus generates $\mathfrak{G}$, and it follows that

$$x^{-1}(\exp A)\,x \in \mathfrak{H} \quad \text{for all } x \in \mathfrak{G},\ A \in \mathfrak{h}.$$

Similarly, as A runs through $\mathfrak{h}$, $\exp A$ runs through a nucleus of $\mathfrak{H}$, which generates $\mathfrak{H}$, therefore $x^{-1}ax \in \mathfrak{H}$ for all $x \in \mathfrak{G}$, $a \in \mathfrak{H}$; this proves that $\mathfrak{H}$ is normal in $\mathfrak{G}$.

Ex. Show that the kernel of the adjoint representation of a connected Lie group $\mathfrak{G}$ is the centre of $\mathfrak{G}$.

CHAPTER VII

THE UNIVERSAL COVERING GROUP

We have seen that two Lie groups which have isomorphic Lie algebras need not be isomorphic themselves, but only locally isomorphic. The question therefore arises of classifying in some way the Lie groups which are locally isomorphic to a given one. We shall not attempt such a classification; however, we shall show that to each connected Lie group $\mathfrak{G}$ there corresponds a uniquely determined Lie group $\tilde{\mathfrak{G}}$, called the *universal covering group* of $\mathfrak{G}$, which is locally isomorphic to $\mathfrak{G}$ and is such that every connected Lie group which is locally isomorphic to $\mathfrak{G}$ is a homomorphic image of $\tilde{\mathfrak{G}}$. Thus if $\mathfrak{g}$ is a given Lie algebra, as soon as we have one connected Lie group corresponding to $\mathfrak{g}$, we can find all such groups by going over to the universal covering group $\tilde{\mathfrak{G}}$ and considering those homomorphic images of $\tilde{\mathfrak{G}}$ which are locally isomorphic to $\tilde{\mathfrak{G}}$.

However, this does not solve the problem of constructing one Lie group corresponding to a given Lie algebra. The usual way of doing this is by means of Ado's theorem. This theorem asserts that every finite dimensional Lie algebra L over R is isomorphic to a subalgebra of $\mathfrak{gl}(r, R)$ for some r. By Theorem 6.7.2 it then follows that there is a connected analytic subgroup of $\mathfrak{GL}(r, R)$ whose Lie algebra is isomorphic to L. This then assures us of the existence of a connected Lie group corresponding to a given Lie algebra. However, Ado's theorem belongs more properly to the theory of Lie algebras and we shall not prove it here.

7.1. Connected and path-connected spaces.

7.1. Connected and path-connected spaces. We recall that a topological space T is said to be *connected*, if

C. 1. *There is no partition of T into two non-empty closed subsets.*

For later use we note the following properties which are equivalent to C. 1 and may therefore also be used to define connectedness:

C. 2. *There is no partition of T into two non-empty open subsets.*

C.3. *Given any partition of T into open sets:*

$$T = \bigcup_{\alpha} O_{\alpha} \quad (O_{\alpha} \cap O_{\beta} = \emptyset \text{ for } \alpha \neq \beta),$$

all except one of the O_{α} must be empty.

C.4. *If T is mapped continuously into a discrete space, then the image of T consists of a single point.*

We establish the equivalence of these four properties by proving the scheme: C.1 $\Leftrightarrow$ C.2 $\Rightarrow$ C.4 $\Rightarrow$ C.3 $\Rightarrow$ C.2. That C.2 is equivalent to C.1 is seen by taking complements. Next, if C.2 holds and ϕ is a continuous mapping of T into a discrete space A, then for any $a \in A$, $\overset{-1}{\phi}(a)$ is open; hence if $\phi(T)$ had more than one element,

$$T = \overset{-1}{\phi}(a) \cup \bigcup_{b \neq a} \overset{-1}{\phi}(b)$$

would be a partition of T into two non-empty open subsets, which contradicts C.2; therefore C.2 implies C.4. Further, C.4 implies C.3, because if $T = \bigcup O_{\alpha}$ is a partition into open sets, then the mapping ϕ defined by the rule

$$\phi(x) = \alpha \quad \text{if } x \in O_{\alpha},$$

is a continuous mapping of T into the set of suffixes $\{\alpha\}$ regarded as a discrete space. By C.4 the image of T consists of a single point, which means that all but one of the O_{α} are empty. Finally, C.2 is a special case of C.3; this completes the proof that C.1–4 are equivalent.

To verify that a space is connected we may therefore use any one of C.1–4.

If T is any topological space, a *path* in T is a continuous mapping of the closed interval $I = [0, 1]$ of real numbers into T. Denote the coordinate in I by s, then the path is described by a continuous function $h(s)$, defined for $0 \leqslant s \leqslant 1$ with values in T. The path is said to go from the *initial point* $h(0)$ to the *final point* $h(1)$; $h(0)$ and $h(1)$ are also called the *end-points* of the path. When the end-points coincide: $h(0) = h(1) = p$, say, the path is said to be a *loop*, or, more precisely, a *loop on* p.† We note that a path in T is not just a point set in T, but a mapping of I into T. Thus if we make a change of coordinate in I we obtain a different path.

† A loop is sometimes also called a *closed path.*

When no confusion is possible we use the same letter, h say, to denote the path and the function $h(s)$ defining it.

With every path h from a point p to a point q of T we can associate a path from q to p, defined by the mapping

$$h'(s) = h(1-s).$$

This path will be denoted by h^{-1}. Clearly we have $(h^{-1})^{-1} = h$. If h and k are paths from p to q and q to r respectively, then the mapping

$$l(s) = \begin{cases} h(2s) & (0 \leqslant s \leqslant \frac{1}{2}), \\ k(2s-1) & (\frac{1}{2} \leqslant s \leqslant 1), \end{cases}$$

is continuous, and so defines a path from p to r, which will be denoted by hk. Further we associate with every point p of T the constant mapping $\quad h(s) = p \quad (0 \leqslant s \leqslant 1)$,

which we denote by O_p. It follows that every point can be joined to itself by a path, and if two points can be joined to a third, they can be joined to each other. More briefly we may say that the relation 'There exists a path from p to q' is reflexive, symmetric and transitive.

We define a space T to be *path-connected*†, if any two points of T can be joined by a path. By what we have just shown it is enough to require that there exist one point of T which can be joined to any other point of T by a path.

It is not difficult to see that a path-connected space must be connected. For if such a space were not connected we should be able to find a continuous mapping ϕ onto a discrete space $(a, b, \ldots)$ with more than one element. Let p and q be points mapping into a and b respectively; by hypothesis p and q can be joined by a path, h say. Then the mapping

$$s \to \phi(h(s))$$

is a continuous mapping of $I = [0, 1]$ into a discrete space with $\phi(h(0)) = a \neq b = \phi(h(1))$, and this contradicts the fact that the closed interval I is connected.‡

† It can be shown that in any Hausdorff space two distinct points which can be joined by a path can also be joined by an arc (i.e. a homeomorph of the closed unit interval). For this reason such a space is often called *arc-connected* or *arc-wise connected*. We prefer the name used in the text as we have no occasion to deal with arcs.

‡ Cf., for example, Bourbaki[2], p. 73.

For general spaces the notions 'connected' and 'path-connected' are not equivalent; however, they are equivalent for manifolds:

THEOREM 7.1.1. *A manifold M is connected if and only if it is path-connected.*

Proof. We have just seen that any path-connected space is connected. Now let M be any connected manifold and for each point $p \in M$ let S_p be the set of points which can be joined to p by a path. Any two such sets S_p, S_q are either disjoint or coincide, for they are just the equivalence classes of the relation 'p can be joined to q by a path'. Since each point of M belongs to some S_p, the different sets S_p define a partition of M. Further, each S_p is open, for if $q \in S_p$ and U is a spherical neighbourhood of q in terms of some chart at q, then every point of U can be joined to q by a path, and so also to p; hence $U \subseteq S_p$, and this proves that S_p is open. Thus we have a partition of M into open sets, and since M is connected, all except one, say S_p, must be empty. This means that every point of M can be joined to p by a path, i.e. M is path-connected, as we wished to prove.

7.2. Homotopy. We now have to consider under what circumstances two paths between the same points may be considered as equivalent. Let p, q be two points of a topological space T and h_1, h_2 two paths from p to q. Then we say that h_1 and h_2 are *homotopic*: $h_1 \sim h_2$, if there exists a function $\phi(s,t)$ of two real variables, defined and continuous in the unit square

$$Q: 0 \leqslant s \leqslant 1, \quad 0 \leqslant t \leqslant 1$$

of the Euclidean plane, with values in T, such that

$$\phi(0,t) = p, \quad \phi(1,t) = q \quad (0 \leqslant t \leqslant 1),$$
$$\phi(s,0) = h_1(s), \quad \phi(s,1) = h_2(s) \quad (0 \leqslant s \leqslant 1).$$

If we interpret t as the time we may think of h_1 as being deformed continuously into h_2, the end-points p, q remaining fixed. This intuitive picture suggests the following terminology: If $h_1 \sim h_2$, then we say that *h_1 can be deformed into h_2 across $\phi(Q)$.*

It is not hard to see that the relation '$\sim$' is reflexive, symmetric and transitive, and therefore defines a division of paths into mutually exclusive classes, called the *homotopy classes*. We give only the proof of the transitivity: Let $h_1 \sim h_2$, $h_2 \sim h_3$, and let $\phi(s,t)$, $\psi(s,t)$ be the functions defining these homotopies, so that

$$\phi(s,1) = h_2(s) = \psi(s,0) \quad (0 \leqslant s \leqslant 1).$$

Then the function $\chi(s,t)$, defined by

$$\chi(s,t) = \begin{cases} \phi(s,2t) & (0 \leqslant t \leqslant \frac{1}{2}), \\ \psi(s,2t-1) & (\frac{1}{2} \leqslant t \leqslant 1), \end{cases}$$

is continuous and this shows that $h_1 \sim h_3$.

The proof may be summarized as follows: If h_1 can be deformed into h_2 and h_2 can be deformed into h_3, then h_1 can be deformed into h_3. This formulation, while not amounting to a proof, does suggest the lines along which a formal proof would run; at the same time it gives an intuitive picture of the situation. We shall therefore on occasions use the looser terminology and leave the task of formulating a rigorous proof to the reader.

The homotopy relation satisfies the following rules relative to the operations on paths introduced in 7.1:

(i) *If $h_1 \sim h_2$, $k_1 \sim k_2$, and if the final point of h_1 coincides with the initial point of k_1—so that $h_1 k_1$ and also $h_2 k_2$ are defined—then $h_1 k_1 \sim h_2 k_2$ and $h_1^{-1} \sim h_2^{-1}$.*

(ii) *If h is any path beginning at p, then $hh^{-1} \sim O_p$.*

(iii) *If h is any path from p to q, then $O_p h \sim hO_q \sim h$.*

(iv) *If h, k, l are any paths such that hk and kl are defined, then $(hk)l$ and $h(kl)$ are defined, and they are homotopic.*

The proof of (i) is immediate. To prove (ii), we put

$$\phi(s,t) = \begin{cases} h((1-t)\,2s) & (0 \leqslant s \leqslant \frac{1}{2}), \\ h(2(1-t)(1-s)) & (\frac{1}{2} \leqslant s \leqslant 1); \end{cases}$$

this defines a homotopy between hh^{-1} and O_p, as is easily verified. In (iii) the homotopy between hO_q and h may be defined by the function

$$\phi(s,t) = \begin{cases} h\left(\dfrac{2s}{1+t}\right) & \left(0 \leqslant s \leqslant \dfrac{1+t}{2}\right), \\ q & \left(\dfrac{1+t}{2} \leqslant s \leqslant 1\right), \end{cases}$$

and the assertion $O_p h \sim h$ is proved similarly. Finally, to prove (iv), we note that k and kl have the same initial point, which must be the final point of h because hk is defined. Hence $h(kl)$ exists, and similarly we see that $(hk)l$ exists. Their homotopy is now defined by the function

$$\phi(s,t)=\begin{cases} h\left(\dfrac{4s}{2-t}\right) & \left(0\leqslant s\leqslant \dfrac{2-t}{4}\right), \\ k(4s+t-2) & \left(\dfrac{2-t}{4}\leqslant s\leqslant \dfrac{3-t}{4}\right), \\ l\left(\dfrac{4s+t-3}{1+t}\right) & \left(\dfrac{3-t}{4}\leqslant s\leqslant 1\right). \end{cases}$$

If we denote the homotopy class containing a given path h by $\{h\}$, it follows from (i) that we can define a multiplication between homotopy classes of paths h, k such that the final point of h coincides with the initial point of k. We simply put

$$\{h\}\{k\}=\{hk\}; \tag{1}$$

then (i) shows that this product is independent of the choice of the representatives h and k in their homotopy classes. In particular, if we limit ourselves to loops on a given point p, then the product of two loops is always defined, and (ii)–(iv) imply that the homotopy classes form a group with respect to the multiplication defined by (1). The unit element of this group is $\{O_p\}$ and the inverse of any class $\{h\}$ is $\{h^{-1}\}$.

A manifold M is said to be *simply connected* if it is connected and if every loop is homotopic to zero. In detail this means: given any point $p\in M$ and any loop h on p, we have $h\sim O_p$. For the applications it is convenient to note that this condition need only be satisfied at a single point:

THEOREM 7.2.1. *Let M be a connected manifold and p any point of M. Then each of the following conditions is necessary and sufficient for M to be simply connected:*

(i) *Any loop on p is homotopic to O_p.*

(ii) *Any two paths from p to the same point are homotopic.*

Proof. We first note that (i) and (ii) are equivalent: (i) is the special case of (ii) obtained by taking one of the paths to be O_p. Conversely, if (i) holds and h_1, h_2 are any two paths from p to q, then $h_1h_2^{-1}$ is a loop on p, hence $h_1h_2^{-1}\sim O_p$ by (i), and so

$$h_2\sim O_ph_2\sim h_1h_2^{-1}h_2\sim h_1O_q\sim h_1.$$

Clearly (i) is satisfied in any simply connected manifold, so t M be a connected manifold satisfying (i), and k any loop on q, ay. Since M is connected, there is a path h from p to q, and kh^{-1} is a path from p to p. By (i), we have $hkh^{-1} \sim O_p$, hence $\sim O_q k O_q \sim h^{-1}hkh^{-1}h \sim h^{-1}O_p h \sim O_q$, i.e. $k \sim O_q$, as was to be roved.

7.3. Covering homomorphisms. A homomorphism of ne topological group G onto another, H, is called a *covering omomorphism*, if it is a local isomorphism. Such a homonorphism is necessarily continuous and open, since it is coninuous and open at the unit element of G. Covering homonorphisms are characterized by the fact that their kernels are liscrete subgroups:

THEOREM 7.3.1. *Let G and H be topological groups and let θ be ı covering homomorphism of G onto H. Then the kernel D of θ is ı discrete normal subgroup of G, and H is topologically isomorphic to G/D. Conversely, if D is a discrete normal subgroup of G then the natural mapping of G onto G/D is a covering homomorphism.*

Proof. Let V be a nucleus of G such that the restriction $\theta \mid V$ is one-one, and denote the kernel of θ by D. Then $V \cap D = \{e\}$, since there can only be one element of V which maps into e^{θ}. Now $V \cap D$ is a nucleus of D in the topology induced by G, and D is therefore discrete. The mapping $\theta \mid V$, being a topological isomorphism, is continuous and open, whence the induced isomorphism of G/D onto H is continuous and open (by Theorem 2.3.1), and therefore a topological isomorphism. The converse is immediate: If D is discrete, then there is a nucleus V of G such that $V \cap D = \{e\}$, and the natural mapping of G onto G/D, when restricted to V, is one-one. Since it is also continuous and open, it is a topological isomorphism of V with its image in G/D. This completes the proof.

In a connected group the kernel of a covering homomorphism has to satisfy a further condition, given by

THEOREM 7.3.2. *Let G be a connected topological group and θ a covering homomorphism of G (onto some other group). Then the kernel of θ is a discrete subgroup of the centre of G.*

Proof. Denote the kernel of θ by D. Since θ is a local isomorphism, D is a discrete normal subgroup of G. The theorem therefore follows from the

LEMMA. *Let G be a connected group and D a discrete normal subgroup of G. Then D is contained in the centre of G.*

To prove the lemma, consider, for any fixed $a \in D$, the mapping

$$x \rightarrow x^{-1}ax \quad (x \in G)$$

of G into D. This is a continuous mapping of a connected space, namely, G, into a discrete space, D, and hence the image consists of a single point. Thus $x^{-1}ax = e^{-1}ae = a$ for all $x \in G$, $a \in D$, which is just the assertion of the lemma.

7.4. Construction of the universal covering group. If $\mathfrak{G}$ and $\overline{\mathfrak{G}}$ are two connected Lie groups, then $\overline{\mathfrak{G}}$ is said to be a *covering group* of $\mathfrak{G}$ if there is a covering homomorphism of $\overline{\mathfrak{G}}$ onto $\mathfrak{G}$. When $\mathfrak{G}$ is given, covering groups of $\mathfrak{G}$ may be constructed as follows:

Let V be an open connected nucleus of $\mathfrak{G}$; take a set $\overline{V} = \{\bar{x}, \bar{y}, \ldots\}$ in one-one correspondence with V and denote by G_0 the abstract group with the elements of $\overline{V}$ as generators and the relations

$$\bar{x}\bar{y} = \bar{z} \text{ whenever } xy = z \text{ for the corresponding elements of } V. \quad (2)$$

The mapping $\bar{x} \rightarrow x$ of $\overline{V}$ into V can be extended to a homomorphism of G_0 into $\mathfrak{G}$ which we denote by ϕ. For every element of G_0 has the form

$$\bar{x}\bar{y} \ldots \bar{z}, \quad \text{where } x, y, \ldots, z \in V; \quad (3)$$

its image belongs to $\mathfrak{G}$ and by (2) is independent of the way in which the element (3) has been expressed in terms of the generators. The restricted mapping $\phi \mid \overline{V}$ is one-one, by construction, and it is an isomorphism between $\overline{V}$ and V. Now V is a local Lie group; if we transfer the analytic structure from V to $\overline{V}$ by means of ϕ, $\overline{V}$ also becomes a local Lie group. Moreover, $\overline{V}$ generates G_0 and so we may, by Theorem 2.7.1, use the analytic structure to define G_0 as a connected Lie group, which we denote by $\mathfrak{G}_V$.

We note that ϕ is a homomorphism of $\mathfrak{G}_V$ into $\mathfrak{G}$ which is analytic in $\overline{V}$, and hence everywhere. The image of $\mathfrak{G}_V$ under ϕ

is a subgroup of $\mathfrak{G}$ containing the nucleus V, and hence its image coincides with $\mathfrak{G}$, because $\mathfrak{G}$ is connected (Theorem 2.4.3). Thus ϕ maps $\mathfrak{G}_V$ onto $\mathfrak{G}$. Finally, ϕ is a local isomorphism since it is one-one on $\bar{V}$ (Theorem 6.3.4). Thus $\mathfrak{G}_V$ is a covering group of $\mathfrak{G}$; it is called the *covering group of order* V of $\mathfrak{G}$. The facts proved about $\mathfrak{G}_V$ may be summed up as

LEMMA 7.4.1. *If $\mathfrak{G}$ is any connected Lie group and V an open connected nucleus of $\mathfrak{G}$, then $\mathfrak{G}_V$, the covering group of order V of $\mathfrak{G}$, is connnected and is generated by a subset $\bar{V}$ which is mapped in a one-one manner onto V by the covering homomorphism between $\mathfrak{G}_V$ and $\mathfrak{G}$.*

Since the nucleus V is connected, $\bar{V}$ may be characterized as the identity component of the inverse image $\overset{-1}{\phi}(V)$. It is also worth noting that the covering group $\mathfrak{G}_V$ depends only on V and not on any part of $\mathfrak{G}$ outside V. This is clear from the construction of $\mathfrak{G}_V$. All that is required is the existence of $\mathfrak{G}$, i.e. in the above construction we need to know that the local Lie group V is embeddable in a Lie group $\mathfrak{G}$.† From this remark we derive

LEMMA 7.4.2. *If $\mathfrak{G}_1$ and $\mathfrak{G}_2$ are connected Lie groups which are locally isomorphic, then they have a common covering group.*

Proof. Let V_i be an open nucleus of $\mathfrak{G}_i$ $(i=1,2)$ such that $V_1 \leftrightarrow V_2$ under a topological isomorphism. If V_1 is taken to be connected, so is V_2; then $\mathfrak{G}_{V_1} \simeq \mathfrak{G}_{V_2}$, and identifying these groups we obtain a covering group of both $\mathfrak{G}_1$ and $\mathfrak{G}_2$, which is what we had to find.

Whenever we have a connected Lie group $\mathfrak{G}$ and a covering group $\bar{\mathfrak{G}}$ we know that there is a homomorphism of $\bar{\mathfrak{G}}$ onto $\mathfrak{G}$ which is one-one on a sufficiently small subset $\bar{U}$ of $\bar{\mathfrak{G}}$; the next problem is to find conditions for a subset $\bar{U}$ of $\bar{\mathfrak{G}}$ to have this property. For the applications it is more convenient to state this condition in terms of the image of $\bar{U}$ in $\mathfrak{G}$.

THEOREM 7.4.3. *Let $\mathfrak{G}$ be a connected Lie group and $\bar{\mathfrak{G}}$ a covering group of $\mathfrak{G}$ with covering homomorphism ϕ. If U is a simply*

† That it is possible to embed *every* local Lie group in a global group follows from the full converse of Lie's third theorem.

connected open subset of $\overline{\mathfrak{G}}$ *and* $\overline{U}$ *is a connected component of th inverse* $\overset{-1}{\phi}(U)$*, then the restriction* $\phi \mid \overline{U}$ *is one-one.*

For the proof of this theorem—as well as for later purposes—we require the theorem on the uniform continuity of continuou functions on a compact metric space; it is convenient to state thi as follows:

LEMMA 7.4.4. *Let* θ *be a continuous mapping of a bounded close subset* A *of* R^n *into a topological space* T*, and suppose that for eac point* p *of* T *a neighbourhood* W_p *has been chosen. Then there exist* $\delta > 0$ *such that any closed spherical neighbourhood of a point of of radius* δ *is mapped into some* W_p *by* θ*.*

Proof. If $a \in A$ and $a^\theta = p$, then by the continuity of θ, $\overset{-1}{\theta}(W_p)$ is a neighbourhood of a. Hence there is some closed spherica neighbourhood of centre a whose image under θ is contained in W_p. *A fortiori* there is a closed spherical neighbourhood of centre a whose image under θ is contained in W_q for some $q \in T$. Denote by δ_a the least upper bound of the radii of such neighbourhoods; by what has been said, $\delta_a > 0$ for each $a \in A$ and we have to prove that the δ_a have a positive lower bound, as a varies over A. Suppose not, and let $a_1, a_2, \ldots$ be a sequence of points such that $\delta_{a_i} \to 0$. By going over to a subsequence if necessary we may suppose that the points a_i converge to a point $a \in A$ (because A is bounded and closed). Now $\delta_a > 0$, and for points a_i at a distance less than $\frac{1}{2}\delta_a$ from a we have $\delta_{a_i} \geqslant \frac{1}{2}\delta_a$, which contradicts the supposition $\delta_{a_i} \to 0$. Thus the lemma is established.†

We can now prove Theorem 7.4.3: By definition of ϕ as a local isomorphism there is an open nucleus $\overline{V}$ of $\overline{\mathfrak{G}}$ which is mapped homeomorphically onto an open nucleus V of $\mathfrak{G}$. Replacing $\overline{V}$ by a smaller nucleus if necessary—e.g. a spherical neighbourhood in terms of some chart—we may suppose that $\overline{V}$ is open connected and such that the restriction of ϕ to $\overline{V}^{-1}\overline{V}$ is still one-one. Let D be the kernel of the homomorphism ϕ. Then D is discrete and $D \cap \overline{V}^{-1}\overline{V} = \{\bar{e}\}$, where $\bar{e}$ is the unit element of $\overline{\mathfrak{G}}$. Hence $\overline{V}D$ is the union of the disjoint open sets $\overline{V}d$ $(d \in D)$, and by translation,

† Cf. Lemma 6.1.1, which is based on the same fact.

$D\bar{x}$ is the union of the disjoint open sets $\overline{V}d\bar{x}\ (d \in D)$, for any $\in \overline{\mathfrak{G}}$. It follows that any connected component of $\overset{-1}{\phi}(Vx)$, where $\in \mathfrak{G}$, has the form $\overline{V}\bar{x}\ (\bar{x} \in \overline{\mathfrak{G}})$.

Now let $\bar{x}$, $\bar{y} \in \overline{U}$ and suppose that $\phi(\bar{x}) = \phi(\bar{y}) = x'$, say. We ave to show that $\bar{x} = \bar{y}$. Since $\overline{U}$ is connected and open, it is a ıanifold, and therefore path-connected; hence we may join $\bar{x}$ and by a path h in $\overline{U}$. Let h' be its image under ϕ; more precisely, $'$ is the path defined by $h'(s) = \phi(h(s))\ (0 \leqslant s \leqslant 1)$. We shall xpress this more briefly by writing $h' = \phi(h)$. Clearly h' is a loop n x'. Since U is simply connected, $h' \sim O_{x'}$. We complete the roof by showing that the deformation of h' to a point can also e carried out on h, so that its end-points must coincide, i.e. $= \bar{y}$.

Consider then the statement $h' \sim O_{x'}$. This means that there is continuous mapping θ of a square Q into U such that one side maps onto h' and the other three onto x'. If $u \in U$, then Vu is a neighbourhood of u, and by the lemma just proved (7.4.4), there exists $\delta > 0$, such that any closed circular neighbourhood of radius δ maps into some Vu. In particular, if we subdivide the square Q into equal squares of side less than $\sqrt{2}\,\delta$, each of the small squares is mapped into some Vu by θ.

Let Q_1 be one of these small squares lying on that side of Q which maps into h'. Its image $\theta(Q_1)$ meets the path h' and hence, if $\theta(Q_1) \subseteq Vu$, then Vu meets h' and therefore some connected component $\overline{V}\bar{u}$ say, of $\overset{-1}{\phi}(Vu)$, meets the path h. The restriction $\phi \mid \overline{V}\bar{u}$ is a homeomorphism between $\overline{V}\bar{u}$ and Vu, and we may therefore deform h' across $\theta(Q_1)$ and at the same time carry out a corresponding deformation of h, so that the relation $h' = \phi(h)$ continues to hold. We can now repeat the process, using any of the small squares adjacent to Q_1. In this way we can shrink h' by deforming it across the images of the small squares, while preserving the relation $h' = \phi(h)$. The process ends when h' has been shrunk to the point x', and h is then a path from $\bar{x}$ to $\bar{y}$ which belongs entirely to $\overset{-1}{\phi}(x')$. But $\overset{-1}{\phi}(x') = \bar{x}D$ is a discrete space, hence h must then reduce to a single point, i.e. $\bar{x} = \bar{y}$. This proves the theorem.

COROLLARY. *If $\overline{\mathfrak{G}}$ is a covering group of a simply connected L group $\mathfrak{G}$, then $\overline{\mathfrak{G}}$ is topologically isomorphic to $\mathfrak{G}$.*

For we need only take $U = \mathfrak{G}$ in Theorem 7.4.3.

Thus in order to obtain a universal covering group of $\mathfrak{G}$ w have to find a simply connected covering group of $\mathfrak{G}$. It existence, which is proved in the next theorem, depends essen tially on the fact that every Lie group has a simply connecte nucleus. To obtain such a nucleus we take a chart at e such tha $e = (0, 0, \ldots, 0)$ and choose a spherical neighbourhood U of e i terms of this chart. Then U is connected and it is simply con nected, for if h is any loop on e, let $h^i(s)$ be the coordinates of th point $h(s)$ in terms of the given chart. Then the function θ define by

$$\theta^i(s, t) = (1 - t)\, h^i(s)$$

provides a homotopy between h and O_e.

THEOREM 7.4.5. *Any connected Lie group $\mathfrak{G}$ has a simpl connected covering group. More precisely, if U is a simply connected open nucleus of $\mathfrak{G}$ and V an open connected nucleus such that $VV \subseteq U$, then the covering group of order V of $\mathfrak{G}$ is simply connected.*

Proof. Let $\mathfrak{G}_V$ be the covering group of order V of $\mathfrak{G}$, with covering homomorphism ϕ of $\mathfrak{G}_V$ onto $\mathfrak{G}$. If $\overline{U}$, $\overline{V}$ are the identity components of $\overset{-1}{\phi}(U)$, $\overset{-1}{\phi}(V)$ respectively, then by Theorem 7.4.3, the restriction $\phi \mid \overline{U}$ is a homeomorphism onto U, and $\overline{V}$ corresponds to V under this mapping, so that $\overline{V} \subseteq \overline{U}$. Moreover, $\overline{V}\,\overline{V}$ is a connected subset of $\overset{-1}{\phi}(U)$ which meets the component $\overline{U}$, so that $\overline{V}\,\overline{V} \subseteq \overline{U}$.

With each product $v = \bar{x}_1 \ldots \bar{x}_n$ of elements of $\overline{V}$ we shall associate a certain homotopy class of paths from the unit element $\bar{e}$ of $\mathfrak{G}_V$ to v, which is defined as follows. We have

$$\begin{aligned} \bar{x}_1 \ldots \bar{x}_i &\in \bar{x}_1 \ldots \bar{x}_{i-1} \overline{V} \\ &\subseteq \bar{x}_1 \ldots \bar{x}_{i-1} \overline{U}. \end{aligned}$$

The set on the right, being obtained by translation from $\overline{U}$, is simply connected. Hence there is a path h_i from $\bar{x}_1 \ldots \bar{x}_{i-1}$ to $\bar{x}_1 \ldots \bar{x}_i$, running entirely within $\bar{x}_1 \ldots \bar{x}_{i-1}\overline{V}$ and determined to ithin homotopy in $\bar{x}_1 \ldots \bar{x}_{i-1} \bar{U}$. Taking $\bar{x}_0$ to be $\bar{e}$, we now form e path $h = h_1 h_2 \ldots h_n$ and denote its homotopy class by

$$\Psi(\bar{x}_1, \ldots, \bar{x}_n).$$

learly h runs from $\bar{e}$ to v, and the class Ψ might seem to depend n the $\bar{x}_i$, but it does not depend on the h_i. If

$$\bar{x}\bar{y} = \bar{z} \tag{4}$$

any relation in $\bar{V}$, then the paths in both the classes $\Psi(\bar{x}, \bar{y})$ nd $\Psi(\bar{z})$ run from $\bar{e}$ to $\bar{z}$ and they may be taken to lie in $\bar{V}\bar{V} \subseteq \bar{U}$. ince $\bar{U}$ is simply connected, it follows that $\Psi(\bar{x}, \bar{y}) = \Psi(\bar{z})$ and, nore generally, the class $\Psi(\bar{x}_1, \ldots, \bar{x}_n)$ is unchanged by a substi-ution of the form $\bar{x}\bar{y} = \bar{z}$, since the change in a representative path akes place in a region of the form $u\bar{U}$ ($u \in \mathfrak{G}_V$). Now the relations 4) are the defining relations of the group $\mathfrak{G}_V$ and therefore

$$\Psi(\bar{x}_1, \ldots, \bar{x}_n) = \Psi(\bar{y}_1, \ldots, \bar{y}_m)$$

vhenever $\bar{x}_1 \ldots \bar{x}_n = \bar{y}_1 \ldots \bar{y}_m$.

In other words, Ψ depends only on the end-point v; we may therefore write it as $\Psi(v)$.

We complete the proof by showing that every path from $\bar{e}$ to v lies in $\Psi(v)$. Then it will follow that all the paths from $\bar{e}$ to v are homotopic, and by Theorem 7.2.1, $\mathfrak{G}_V$ must be simply connected. Let then h be any path from $\bar{e}$ to v. Choose a nucleus $\bar{W}$ of $\mathfrak{G}_V$ such that $\bar{W}\bar{W}^{-1} \subseteq \bar{V}$, then $u\bar{W}$ is a neighbourhood of u for any $u \in \mathfrak{G}_V$, and by Lemma 7.4.4 we can subdivide the interval [0, 1] into subintervals so small that each is mapped by h into some set $u\bar{W}$. Let us denote these closed subintervals by

$$I_i \quad (i = 1, \ldots, n)$$

and their images under h by I'_i, so that

$$I'_i \subseteq v_i \bar{W}, \quad \text{where } v_i \in \mathfrak{G}_V, \tag{5}$$

say. Increasing the number of intervals by one or two if necessary, we may even suppose that $v_0 = \bar{e}$, $v_n = v$. Let u_i be the image of the left-hand end-point of I_i, then $u_i \in I'_{i-1} \cap I'_i$ ($i = 1, \ldots, n$). By (5),

$$u_i = v_{i-1} w = v_i w', \quad \text{where } w, w' \in \bar{W},$$

therefore if we put $\bar{x}_i = v_{i-1}^{-1} v_i$, we have

$$\bar{x}_i = v_{i-1}^{-1} v_i = w w'^{-1} \in \bar{W}\bar{W}^{-1} \subseteq \bar{V}.$$

Thus $$v = v_0(v_0^{-1}v_1)(v_1^{-1}v_2)\dots(v_{n-1}^{-1}v_n)$$
$$= \bar{x}_1\bar{x}_2\dots\bar{x}_n.$$

This shows that $h \in \Psi(v)$. Since h was *any* path from $\bar{e}$ to v, these paths are all homotopic and so $\mathfrak{G}_V$ is simply connected (Theorem 7.2.1), as we wished to prove.

Now it is an easy matter to prove the existence of a 'universal covering group for any connected Lie group.

THEOREM 7.4.6. *Let $\mathfrak{G}$ be a connected Lie group. Then there exists a connected Lie group $\tilde{\mathfrak{G}}$ such that*

(i) *there is a covering homomorphism ϕ of $\tilde{\mathfrak{G}}$ onto $\mathfrak{G}$,*

(ii) *if θ is any topological isomorphism between a nucleus of $\mathfrak{G}$ and a nucleus of a second connected Lie group $\mathfrak{G}_1$, then there is a covering homomorphism $\tilde{\theta}$ of $\tilde{\mathfrak{G}}$ onto $\mathfrak{G}_1$ such that*

$$x^{\tilde{\theta}} = x^{\phi\theta} \quad \textit{for } x \textit{ in some nucleus of } \tilde{\mathfrak{G}}.$$

The group $\tilde{\mathfrak{G}}$ is determined to within analytic isomorphism by $\mathfrak{G}$. More precisely, if $\tilde{\mathfrak{G}}'$ also satisfies (i) *and* (ii), *with covering homomorphism ϕ', then there is an analytic isomorphism α between $\tilde{\mathfrak{G}}$ and $\tilde{\mathfrak{G}}'$ such that*

$$x^{\phi} = x^{\alpha\phi'} \quad \textit{for all } x \in \tilde{\mathfrak{G}}.$$

The properties (i) and (ii) may be expressed more briefly by saying that (i) $\tilde{\mathfrak{G}}$ covers $\mathfrak{G}$, and (ii) $\tilde{\mathfrak{G}}$ covers the identity component of any group locally isomorphic to $\mathfrak{G}$. Since any covering group of $\mathfrak{G}$ is connected and locally isomorphic to $\mathfrak{G}$ it follows that $\tilde{\mathfrak{G}}$ covers every covering group of $\mathfrak{G}$. For this reason $\tilde{\mathfrak{G}}$ is called the *universal covering group* of $\mathfrak{G}$.

To prove the theorem we take a simply connected open nucleus U of $\mathfrak{G}$, e.g. a spherical neighbourhood in terms of some chart, and an open connected nucleus V such that $VV \subseteq U$. By Theorem 7.4.5, the covering group $\mathfrak{G}_V$ of order V is then simply connected. We shall show that $\mathfrak{G}_V$ has the required properties. The property (i) holds by Lemma 7.4.1. To prove (ii), let $\mathfrak{G}$ and $\mathfrak{G}_1$ be locally isomorphic by a mapping θ: $W \to W_1$, where W, W_1 are open nuclei of $\mathfrak{G}$, $\mathfrak{G}_1$ respectively. We may suppose $W \subseteq V$—replacing W by $W \cap V$ if necessary—and we then have the following topological isomorphisms:

$$\theta: W \to W_1, \quad \phi: \overline{W} \to W,$$

here $\overline{W} = \overset{-1}{\phi}(W) \cap \overline{V}$. Clearly the mapping $\phi\theta$ is a topological omorphism between $\overline{W}$ and W_1; by taking the covering groups $_{\overline{W}}$, $\mathfrak{G}_{W_1}$ of $\mathfrak{G}_V$ and $\mathfrak{G}_1$ respectively, we find that $\mathfrak{G}_{\overline{W}} \simeq \mathfrak{G}_{W_1}$. ow $\mathfrak{G}_V$ is simply connected and therefore topologically iso-orphic to $\mathfrak{G}_{\overline{W}}$ (by the Corollary to Theorem 7.4.3). Hence $\mathfrak{G}_V$ also topologically isomorphic to $\mathfrak{G}_{W_1}$, and identifying these roups, we find that $\mathfrak{G}_V$ becomes a covering group of $\mathfrak{G}_1$. If $\tilde{\theta}$ is he covering homomorphism of $\mathfrak{G}_V$ onto $\mathfrak{G}_1$, then $\tilde{\theta}$ agrees ith $\phi\theta$ on $\overline{W}$ and hence $\mathfrak{G}_V$ also satisfies (ii).

If $\tilde{\mathfrak{G}}$, $\tilde{\mathfrak{G}}'$ both satisfy (i) and (ii), with covering homomor-hisms ϕ, ϕ' respectively, then ϕ is one-one in some open nucleus $\tilde{V}$ of $\tilde{\mathfrak{G}}$. Therefore $\overset{-1}{\phi}$ is a topological isomorphism of $\phi(\tilde{V})$ onto $\tilde{V}$, and since $\tilde{\mathfrak{G}}'$ satisfies (ii), there is a covering homomorphism β of $\tilde{\mathfrak{G}}'$ into $\tilde{\mathfrak{G}}$ which satisfies

$$\beta = \phi'\overset{-1}{\phi} \quad \text{in some nucleus of } \tilde{\mathfrak{G}}'.$$

By symmetry, there is a covering homomorphism α of $\tilde{\mathfrak{G}}$ into $\tilde{\mathfrak{G}}'$ such that

$$\alpha = \phi\overset{-1}{\phi'} \quad \text{in some nucleus of } \tilde{\mathfrak{G}}.$$

Now the mapping $\alpha\beta$, when restricted to a suitable nucleus $\tilde{W}$ of $\tilde{\mathfrak{G}}$, reduces to the identity mapping and the set of elements left fixed by $\alpha\beta$ therefore contains $\tilde{W}$; but this set clearly forms a subgroup, and so coincides with $\tilde{\mathfrak{G}}$, i.e. $\alpha\beta$ is the identity mapping on $\tilde{\mathfrak{G}}$. Similarly, $\beta\alpha$ is the identity mapping on $\tilde{\mathfrak{G}}'$, hence α and β are inverses of each other. Thus $\tilde{\mathfrak{G}}$ and $\tilde{\mathfrak{G}}'$ are topologically, and hence analytically, isomorphic; the proof is now complete.

Ex. 1. Every Abelian connected Lie group of dimension n can be written as a direct product $\mathfrak{R}^k \times \mathfrak{T}^{n-k}$ for some k satisfying $0 \leqslant k \leqslant n$.

Ex. 2. If $\mathfrak{G}$ and $\mathfrak{H}$ are any Lie groups and if a continuous mapping ϕ of a nucleus of $\mathfrak{G}$ into $\mathfrak{H}$ is defined satisfying

$$(xy)^\phi = x^\phi y^\phi$$

whenever both sides are defined, then ϕ can be extended to a homomorphism of $\mathfrak{G}$ into $\mathfrak{H}$, provided that $\mathfrak{G}$ is simply connected (cf. Lemma 2.9.2).

APPENDIX

In Chapter V we used a theorem on the integration of a system of total differential equations. The usual proof proceeds by a reduction to the corresponding theorem on ordinary differential equations.† As we also need the fact that the solution is analytic in the independent variables and the initial values, we give a different proof which does not assume any theorems on differential equations. It is based on Cauchy's method of majorants, adapted to the case of several independent variables.

Throughout this appendix we shall write all suffixes as subscripts and reserve superscripts to denote exponents. We also write out all summation signs.

THEOREM A 1. *Let nr functions $\Phi_{i\alpha}(t,x)$ of $r+n$ variables t_α, x_i ($\alpha=1,\ldots,r$; $i=1,\ldots,n$) be given, which are analytic at $t_\alpha=0$, $x_i=0$. Then the equations*

$$\frac{\partial x_i}{\partial t_\alpha}=\Phi_{i\alpha}(t,x) \quad (x_i=a_i \text{ when } t_\alpha=0), \tag{1}$$

have a unique solution $\quad x_i=\phi_i(t,a),$ (2)

analytic at $t_\alpha=0$, $a_i=0$, provided that

$$D_\alpha\Phi_{i\beta}=D_\beta\Phi_{i\alpha} \quad \textit{for } t_\alpha \textit{ and } x_i \textit{ near } 0, \tag{3}$$

where

$$D_\alpha=\frac{\partial}{\partial t_\alpha}+\sum_j \Phi_{j\alpha}(t,x)\frac{\partial}{\partial x_j}. \tag{4}$$

We note that the condition (3) is vacuous when $r=1$.

Proof. From (3) it follows by a simple computation that

$$D_\alpha D_\beta f=D_\beta D_\alpha f$$

for any analytic function f in the x's and t's. We calculate a formal solution (2) of (1) by putting

$$\phi_i(0,a)=a_i, \tag{5}$$

$$\left[\frac{k!}{k_1!\ldots k_r!}\frac{\partial^k}{\partial t_1^{k_1}\ldots\partial t_r^{k_r}}\phi_i(t,a)\right]_{t=0}=\left[\sum_{(\gamma)} D_{\gamma_1}\ldots D_{\gamma_{k-1}}\Phi_{i\gamma_k}\right]_{x=a,\,t=0}, \tag{6}$$

† Cf., for example, Carathéodory, *Variationsrechnung*, pp. 23 ff.

here the sum on the right is extended over all the $\dfrac{k!}{k_1!\dots k_r!}$ ermutations $(\gamma_1 \dots \gamma_k)$ of $(1^{k_1} \dots r^{k_r})$. By (3) and the comutativity of the D's it follows that (6) is equivalent to

$$\left[\frac{\partial^k}{\partial t_1^{k_1}\dots\partial t_r^{k_r}}\phi_i(t,a)\right]_{t=0} = [D_1^{k_1}\dots D_s^{k_s-1}\Phi_{is}]_{x=a,\,t=0},$$

here k_s is the last non-vanishing k_σ.

Regarded as power series in $t_1, \dots, t_r$ only, $\partial\phi_i/\partial t_\alpha$ and $\Phi_{i\alpha}(t,\phi)$ ave the same Taylor coefficients, these coefficients being power eries in the a_j; therefore the $\phi_i(t,a)$ are a solution of (1), provided hat they are analytic at $t_\alpha = 0$, $a_j = 0$. Moreover, any solution of 1) must satisfy (5) and (6) and is thus completely determined, ince these equations enable us to calculate all its Taylor oefficients. To complete the proof we need only show then that he $\phi_i(t,a)$ are analytic at $t_\alpha = 0$, $a_j = 0$. This we do by finding a ommon majorant for the ϕ's.

Let δ and η be positive numbers, so chosen that for $|t_\beta| \leqslant \delta$ nd $|x_j| \leqslant \eta$,

$$|\Phi_{i\alpha}(t,x)| \leqslant M \quad (\alpha = 1, \dots, r;\ i = 1, \dots, n),$$

and write

$$\Psi(t_1, \dots, t_r, x_1, \dots, x_n) = M \prod_\alpha \left(1 - \frac{t_\alpha}{\delta}\right)^{-1} \prod_i \left(1 - \frac{x_i}{\eta}\right)^{-1}.$$

Then $\Psi(t,x)$ majorizes each $\Phi_{i\alpha}$ with respect to both the t's and the x's. Hence, if we determine a function $\psi(t, \dots, t_r, a)$ by

$$\psi(0,a) = a,$$

$$\left[\frac{k!}{k_1!\dots k_r!}\frac{\partial^k}{\partial t_1^{k_1}\dots\partial t_r^{k_r}}\psi(t,a)\right]_{t=0} = [\sum_{(\gamma)}\nabla_{\gamma_1}\cdots\nabla_{\gamma_{k-1}}\Psi(t,x)]_{x=a,\,t=0}, \tag{7}$$

where $\nabla_\alpha = \dfrac{\partial}{\partial t_\alpha} + \Psi(t,x)\sum_j \dfrac{\partial}{\partial x_j}$, then $\psi(t,a)$ majorizes each $\phi_i(t,a)$. We complete the proof by showing:

There exists $\delta_1 > 0$ such that the power series for $\psi(t,a)$ converges for $|t_\alpha| < \delta_1$ and $|a| < \frac{1}{2}\eta$.

Write $t_\alpha = \lambda_\alpha u$ and put

$$\theta(u, \lambda_1, \dots, \lambda_r, v) = \psi(\lambda_1 u, \dots, \lambda_r u, v),$$
$$\Theta(u, \lambda_1, \dots, \lambda_r, x_1, \dots, x_n) = \Psi(\lambda_1 u, \dots, \lambda_r u, x_1, \dots, x_n).$$

Clearly we have $\dfrac{\partial\theta}{\partial u}=\sum_{\alpha}\lambda_\alpha\dfrac{\partial\psi}{\partial t_\alpha}$, and using (7), we find

$$\left[\frac{\partial^k}{\partial u^k}\theta(u,\lambda,a)\right]_{u=0}$$

$$=\left[\sum_{k_1+\ldots+k_r=k}\frac{k!}{k_1!\ldots k_r!}\lambda_1^{k_1}\ldots\lambda_r^{k_r}\frac{\partial^k}{\partial t_1^{k_1}\ldots\partial t_r^{k_r}}\psi(t,a)\right]_{t=}$$

$$=[\sum_{k_1+\ldots+k_r=k}\lambda_1^{k_1}\ldots\lambda_r^{k_r}\sum_{(\gamma)}\nabla_{\gamma_1}\ldots\nabla_{\gamma_{k-1}}\Psi(t,x)]_{t=0,\,x=a},$$

$$=[(\Sigma\lambda_\alpha)(\Sigma\lambda_\beta\nabla_\beta)^{k-1}\Psi(t,x)]_{t=0,\,x=a}.$$

We shall write this in the form

$$\left[\frac{\partial^k}{\partial u^k}\theta(u,\lambda,a)\right]_{u=0}=\left[(\Sigma\lambda_\alpha)\frac{D^{k-1}}{Du^{k-1}}\Theta(u,\lambda,x)\right]_{u=0,\,x=a}, \quad (8)$$

where $D\Theta/Du$ stands for $\Sigma\lambda_\alpha\nabla_\alpha\Psi$ and analogously for higher derivatives.

When we put $\lambda_\alpha=1$ $(\alpha=1,\ldots,r)$ and $x_i=v$ $(i=1,\ldots,n)$ in $\Theta(u,\lambda,x)$, we obtain

$$\Theta(u,1,v)=\frac{M}{\left(1-\frac{u}{\delta}\right)^r\left(1-\frac{v}{\eta}\right)^n}$$

$$=\Omega(u,v),\quad \text{say},$$

and
$$\frac{D\Omega}{Du}=\left[\frac{D\Theta}{Du}\right]_{x_i=v,\,\lambda_\alpha=1}=\frac{\partial\Omega}{\partial u}+r\Omega\frac{\partial\Omega}{\partial v}.$$

Equation (8) now becomes

$$\left[\frac{\partial^k}{\partial u^k}\theta(u,1,v)\right]_{u=0}=\left[r\frac{D^{k-1}}{Du^{k-1}}\Omega\right]_{u=0}.$$

It follows that $\theta(u,1,a)$ is the solution of

$$\frac{dv}{du}=r\Omega(u,v)\quad (v=a \text{ when } u=0).$$

But this is simply the equation

$$\frac{dv}{du}=rM\left(1-\frac{u}{\delta}\right)^{-r}\left(1-\frac{v}{\eta}\right)^{-n};$$

s solution is given by

$$-\frac{v}{\eta}=\begin{cases}\left(1-\frac{a}{\eta}\right)\left[1-C\left(1-\frac{a}{\eta}\right)^{-n-1}\left(1-\left(1-\frac{u}{\delta}\right)^{1-r}\right)\right]^{1/(n+1)} & (r\neq 1),\\ \left(1-\frac{a}{\eta}\right)\left[1-C\left(1-\frac{a}{\eta}\right)^{-n-1}\log\left(1-\frac{u}{\delta}\right)\right]^{1/(n+1)} & (r=1),\end{cases}$$

here C is a certain constant, whose value does not matter here. f a satisfies $|a|\leqslant\frac{1}{2}\eta$, we can determine $\delta_1>0$ such that the olution converges for $|u|<\delta_1$, $|a|\leqslant\frac{1}{2}\eta$. Hence $\theta(u,\lambda,a)$ conerges for $|u|<\delta_1$, $|\lambda_\alpha|\leqslant 1$, $|a|\leqslant\frac{1}{2}\eta$; therefore $\psi(t,a)$ converges or $|t_\alpha|<\delta_1$ and $|a|\leqslant\frac{1}{2}\eta$, and the proof is complete.

The theorem may be stated more concisely if we use differential orms. Let us write

$$\Xi_i=dx_i-\sum_\alpha \Phi_{i\alpha}(t,x)\,dt_\alpha,$$

hen, regarding the dt's and dx's as independent variables elated by the equations

$$\Xi_i=0, \tag{9}$$

we find that

$$\begin{aligned} d\Xi_i &= -\sum_\alpha d\Phi_{i\alpha}\wedge dt_\alpha \\ &= -\sum_{j\alpha}\frac{\partial\Phi_{i\alpha}}{\partial x_j}dx_j\wedge dt_\alpha-\sum_{\alpha\beta}\frac{\partial\Phi_{i\alpha}}{\partial t_\beta}dt_\beta\wedge dt_\alpha.\end{aligned}$$

If we use (9) to express dx_i in terms of the dt's, this becomes

$$d\Xi_i=-\sum_{\alpha\beta}\left(\sum_j\frac{\partial\Phi_{i\alpha}}{\partial x_j}\Phi_{j\beta}+\frac{\partial\Phi_{i\alpha}}{\partial t_\beta}\right)dt_\beta\wedge dt_\alpha.$$

Using again D_α defined by (4), we can write the coefficient of $dt_\beta\wedge dt_\alpha$ as $D_\beta\Phi_{i\alpha}-D_\alpha\Phi_{i\beta}$; the vanishing of these expressions is just the condition (3) of Theorem A 1. We may therefore restate this theorem as

THEOREM A 1′. *Let nr functions $\Phi_{i\alpha}(t,x)$ of $r+n$ variables t_α, x_i be given which are analytic at* 0. *Then the equations*

$$dx_i=\sum_\alpha \Phi_{i\alpha}(t,x)\,dt_\alpha \quad (x_i=a_i \text{ when } t_\alpha=0), \tag{10}$$

have a unique solution $x_i=\phi_i(t,a)$, analytic at 0, *provided that in some neighbourhood of* 0,

$$d(\Sigma\Phi_{i\alpha}dt_\alpha)=0,$$

identically in the dt_α, when dx_i is expressed in terms of the dt's by (10).

Finally, we prove two facts about analytic functions which have also been used in the course of the text, viz. the implicit function theorem and the fact that a matrix of analytic functions has an analytic inverse whenever its determinant is different from zero.

THEOREM A 2. *If* $\psi_{ij}(x)$ *are* k^2 *functions of n variables* $x_1, \ldots, x_n$, *analytic at* $x_i = 0$, *and such that* $\Delta(0) = \det \psi_{ij}(0) \neq 0$, *then the matrix* $(\psi_{ij}(x))$ *has an inverse* $(\check{\psi}_{ij}(x))$, *whose elements are analytic functions at* $x_i = 0$.

Proof. Since $\Delta(x)$ is $\neq 0$ when $x_i = 0$, it is $\neq 0$ in some neighbourhood of 0, and we can calculate the inverse matrix by the usual rules. Its elements are polynomials in the ψ's, divided by the determinant $\Delta(x)$. Now polynomials in analytic functions are clearly analytic, and the theorem will follow if we can prove the

LEMMA. *If* $f(x)$ *is a function of* $x_1, \ldots, x_n$, *analytic at* $x_i = 0$, *and if* $f(0) \neq 0$, *then* $1/f(x)$ *exists and is analytic at* $x_i = 0$.

To prove the lemma we write $f(x)$ as

$$f(x) = c(1 - \Sigma \gamma_{r_1 \ldots r_n} x_1^{r_1} \ldots x_n^{r_n})$$
$$= c(1 - h(x)) \quad \text{say,}$$

where $c = f(0) \neq 0$; $h(x)$ is also analytic at 0 and hence the series for $h(x)$ is convergent for $|x_i| < \rho$ say. Since $h(0) = 0$, we can, by continuity, find $\delta (0 < \delta \leqslant \rho)$ such that

$$\Sigma \, | \gamma_{r_1 \ldots r_n} x_1^{r_1} \ldots x_n^{r_n} | < \tfrac{1}{2} \quad (|x_i| < \delta). \tag{11}$$

Thus for $|x_i| < \delta$ the series for $h(x)$ is absolutely convergent and we may write

$$[h(x)]^k = \sum_{l=1}^{\infty} h_{kl}(x),$$

where $h_{kl}(x)$ denotes the terms of $[h(x)]^k$ which are homogeneous of degree l in the x's. By (11) the series

$$[h(x)]^k = \sum_l h_{kl}(x)$$

is absolutely convergent for $|x_i| < \delta$, and so we may interchange the summations.† The function $g(x) = \frac{1}{c} \sum_k [h(x)]^k$ is therefore

† Cf. Titchmarsh, *Theory of Functions*, p. 31.

nalytic at 0, and it clearly satisfies $f(x)\,g(x)=1$. This proves the mma.

Applying the lemma, we find that $1/\Delta(x)$ is analytic at 0, and ence the elements $\psi_{ij}(x)$ of the inverse matrix are analytic at $_i=0$, as we wished to show.

THEOREM A3. *Let $\chi_i(y)$ be n functions of ν variables $y_1, \ldots, y_\nu$, nalytic at $y_\alpha=0$, and suppose that the matrix $(\partial\chi_i/\partial y_\alpha)_{y=0}$ has ank ν. Then $\nu \leqslant n$ and the equations*

$$x_i=\chi_i(y) \tag{12}$$

an be solved near $y_\alpha=0$ for $y_1, \ldots, y_\nu$ in terms of ν of the x's:

$$y_\alpha=\psi_\alpha(x_{i_1}, \ldots, x_{i_\nu}), \tag{13}$$

vhere the ψ's are analytic functions of their arguments at

$$x_i=\chi_i(0).$$

Conversely, if the equations (12) *have such a solution, then*

$$\left(\frac{\partial\chi_i}{\partial y_\alpha}\right)_{y=0}$$

as the rank ν.

Proof. Suppose that $(\partial\chi_i/\partial y_\alpha)_{y=0}$ is of rank ν. From (12) we derive

$$dx_i=\Sigma\frac{\partial\chi_i}{\partial y_\alpha}\,dy_\alpha; \tag{14}$$

by hypothesis the rank of this system is ν at the point $y_\alpha=0$. Since there are n equations, we must have $\nu \leqslant n$, and since there are ν variables dy_α, the rank can never exceed ν. It follows that the rank is ν in some neighbourhood of the origin, and we can solve ν of the equations (14) for the dy's. Renumbering the x's, we may suppose that the first ν equations can be solved:

$$dy_\alpha=\sum_\beta \Phi_{\alpha\beta}(y)\,dx_\beta. \tag{15}$$

Now we can conclude the proof as in Theorem 5.2.3: Let us write

$$X_i=dx_i-\sum_\alpha\frac{\partial\chi_i}{\partial y_\alpha}\,dy_\alpha, \quad Y_\alpha=dy_\alpha-\sum_\beta \Phi_{\alpha\beta}(y)\,dx_\beta;$$

since (15) was obtained by solving the first ν equations (14), we have

$$Y_\alpha=\Sigma T_{\alpha\beta}X_\beta, \tag{16}$$

where $T_{\alpha\beta}(y)$ is a non-singular matrix with analytic elements. Therefore, under the hypothesis $X_i=0$, in

$$dY_\alpha = \sum_\beta dT_{\alpha\beta} \wedge X_\beta + \sum_\beta T_{\alpha\beta}\, dX_\beta,$$

the first sum on the right vanishes because $X_i=0$, and the second sum vanishes by the integrability of the equations (14), hence $dY_\alpha=0$ under the hypothesis $X_i=0$. But the variables $x_{\nu+1}, \ldots, x_n$ do not occur in Y_α or dY_α, so that we only require $X_\beta=0$, which by (16), is equivalent to $Y_\alpha=0$. Thus

$$dY_\alpha=0 \text{ under the hypothesis } Y_\beta=0.$$

This is exactly the integrability condition of Theorem A 1′, and we can therefore integrate the equations (15) with the initial conditions $y_\alpha=0$, when $x_\beta=\chi_\beta(0)$, obtaining a solution of the form

$$y_\alpha=\psi_\alpha(x_1, \ldots, x_\nu). \tag{17}$$

Now consider the functions $\chi_\alpha(\psi_1(x), \ldots, \psi_\nu(x))$ near the point $x_i=\chi_i(0)$. We have

$$\begin{aligned} d\chi_\alpha &= \Sigma \frac{\partial \chi_\alpha}{\partial y_\beta} \frac{\partial \psi_\beta}{\partial x_\gamma} dx_\gamma \\ &= \Sigma \frac{\partial x_\alpha}{\partial y_\beta} \Phi_{\beta\gamma}(y)\, dx_\gamma \\ &= dx_\alpha \end{aligned}$$

by (14) and (15). Hence, by integration,

$$x_\alpha=\chi_\alpha(\psi_1(x), \ldots, \psi_\nu(x)),$$

and similarly $$y_\alpha=\psi_\alpha(\chi_1(y), \ldots, \chi_\nu(y)),$$

whence (17) is the required solution.

Conversely, if the equations (12) have a solution of the form (17), then

$$dy_\alpha=\Sigma \frac{\partial \psi_\alpha}{\partial x_i} dx_i, \tag{18}$$

and since the y's are independent variables, the dy's are linearly independent, whence, by (18), there are at least ν linearly independent dx's. Since these equations (18) are equivalent to (14), the rank of the matrix $(\partial\chi_i/\partial y_\alpha)$ near $y_\alpha=0$ is at least ν, and cannot be greater than ν, because the matrix has only ν olumns. This completes the proof of the theorem.

If the functions χ_i depend on another set of variables $z_1, \ldots, z_\rho$ esides the y's, and are analytic functions in the y's and z's, hen the functions ψ appearing in the solution (13) are clearly nalytic in the x's and z's. Applying Theorem A 3 in this case, ith $\rho = \nu = n$, $x_i = 0$, we obtain (after putting x, y, ϕ for y, z, χ espectively),

THEOREM A 4. *If $\phi_i(x, y)$ are n functions of $2n$ variables $x_1, \ldots, x_n$, $_1, \ldots, y_n$, analytic at $x_i = y_i = 0$, and if $|\partial\phi_i/\partial y_j|_0 \neq 0$, then the quations*

$$\phi_i(x, y) = 0$$

an be solved near $y_i = 0$ for $y_1, \ldots, y_n$ in terms of $x_1, \ldots, x_n$:

$$y_i = \theta_i(x_1, \ldots, x_n),$$

here the θ's are analytic functions of their arguments at $x_i = 0$.

BIBLIOGRAPHY

Besides the works referred to in the text, a brief selection of books on the subject is included.

[1] BOURBAKI, N. *Topologie Générale* (Paris, 1940, 1951), chs. I, II.

[2] BOURBAKI, N. *Topologie Générale* (Paris, 2e éd. 1951), chs. III, IV.

[3] CARTAN, E. *La théorie des groupes finis et continus et l'Analysis situs* (Paris, 1930).

[4] CARTAN, E. *La théorie des groupes finis et continus et la géométrie différentielle*... (Paris, 1937, 1951).

[5] CHEVALLEY, C. *Theory of Lie groups, I* (Princeton, 1946).

[6] CHEVALLEY, C. *Théorie des groupes de Lie. II. Groupes algébriques* (Paris, 1951).

[7] KOWALEWSKI, G. *Einführung in die Theorie der kontinuierlichen Gruppen* (Leipzig, 1931; New York, 1950).

[8] LIE, S. *Theorie der Transformationsgruppen*, u. Mitw. v. F. Engel (Leipzig, 1888–93, 1930).

[9] LIE, S. *Vorlesungen über continuierliche Gruppen*, bearbeitet von G. Scheffers (Leipzig, 1893).

[10] MALCEV, A. I. Sur les groupes locaux et complets. *Doklady Akad. Nauk*, 32 (1941), 606–8.

[11] MAYER, W. and THOMAS, T. Y. Foundations of the theory of Lie groups. *Ann Math.* 36 (1935), 770–822.

[12] MONTGOMERY, D. and ZIPPIN, L. *Topological transformation groups* (New York, 1955).

[13] PONTRJAGIN, L. *Topological groups* (Princeton, 1939, 1946).

LIST OF SYMBOLS

ymbols frequently used in the text, together with their signi-cance and the pages on which they are defined or first used.

R	real numbers	*page* 4
$\mathfrak{R}$	real numbers, regarded as a Lie group	8
T	torus (real numbers mod 1)	8
$\mathfrak{T}$	torus, regarded as a Lie group	9

Vith an analytic manifold $\mathfrak{M}$ (and any point p of $\mathfrak{M}$) are ssociated

$\mathscr{A}$ ($\mathscr{A}_p$)	set of analytic functions in $\mathfrak{M}$ (defined at p)	10
$\mathfrak{L}_p$	space of tangent vectors at p	12
$\mathfrak{L}_p^*$	space of differentials at p (dual of $\mathfrak{L}_p$)	14
$\mathfrak{L}$	set of analytic infinitesimal transformations	17
$\mathfrak{L}^*$	set of (linear) analytic differential forms	18
$\mathfrak{C}$	algebra of analytic differential forms	84

Vith a Lie group $\mathfrak{G}$ are associated

ρ, λ	right and left translations	31
ϕ^i	composition functions (in a given chart)	44
ψ_i^j	transformation functions (in a given chart)	63
$\Lambda(\mathfrak{G})$ or $\mathfrak{g}$	Lie algebra of $\mathfrak{G}$ (set of infinitesimal right translations)	62
Λ^*	set of linear MC-forms	89
$\mathfrak{E}$	exterior algebra on Λ^*; Grassmann algebra of the Lie algebra of $\mathfrak{G}$	90

INDEX OF DEFINITIONS

The numbers refer to pages. Single letters are given in the list on p. 163.